Gravitational Few-Body Dynamics

Using numerical integration, it is possible to predict the individual motions of a group of a few celestial objects interacting with each other gravitationally. In this introduction to the few-body problem, a key figure in developing more efficient methods over the past few decades summarizes and explains them, covering both basic analytical formulations and numerical methods.

The mathematics required for celestial mechanics and stellar dynamics is explained, starting with two-body motion and progressing through classical methods for planetary system dynamics. This first part of the book can be used as a short course on celestial mechanics.

The second part develops the contemporary methods for which the author is renowned – symplectic integration and various methods of regularization. This volume explains the methodology of the subject for graduate students and researchers in celestial mechanics and astronomical dynamics with an interest in few-body dynamics and the regularization of the equations of motion.

SEPPO MIKKOLA is a senior lecturer at the University of Turku and staff member at Tuorla Observatory. He has made important contributions to the regularization of equations of N-body motion. Since he invented 'algorithmic regularization' of few-body system dynamics, it has become the foundation of many simulations worldwide.

Gravitational Few-Body Dynamics

A Numerical Approach

Seppo Mikkola
University of Turku, Finland

CAMBRIDGE
UNIVERSITY PRESS

CAMBRIDGE
UNIVERSITY PRESS

University Printing House, Cambridge CB2 8BS, United Kingdom

One Liberty Plaza, 20th Floor, New York, NY 10006, USA

477 Williamstown Road, Port Melbourne, VIC 3207, Australia

314–321, 3rd Floor, Plot 3, Splendor Forum, Jasola District Centre,
New Delhi – 110025, India

79 Anson Road, #06–04/06, Singapore 079906

Cambridge University Press is part of the University of Cambridge.

It furthers the University's mission by disseminating knowledge in the pursuit of education, learning, and research at the highest international levels of excellence.

www.cambridge.org
Information on this title: www.cambridge.org/9781108491297
DOI: 10.1017/9781108868105

First published 2020

Printed in the United Kingdom by TJ International Ltd, Padstow, Cornwall

A catalogue record for this publication is available from the British Library.

ISBN 978-1-108-49129-7 Hardback

Additional resources for this publication at www.cambridge.org/mikkola

Dedicated to my wife Elisa,
my daughters Virpi and Juulia, and
grand-daughters Minja, Siina and Ronja.

Contents

	Preface	*page* xi
	Introduction	**1**
1	**The Problems**	**3**
1.1	Introductory Presentation of the Problems	3
1.2	Historical (a Point of View of the Author)	10
2	**Two-Body Motion**	**15**
2.1	Solution of Equations of Motion	16
2.2	Orbital Elements	19
2.3	Solving the Basic Kepler's Equation	23
2.4	Two-Body Motion in Short Steps	27
2.5	Evaluation of c-Functions	31
2.6	General Algorithms for c-Functions	32
3	**Analytical Tools**	**35**
3.1	The Lagrangian	35
3.2	The Hamiltonian	36
3.3	Canonical Transformations	37
3.4	Time-Dependent Hamiltonian	39
3.5	Time-Dependent Point Transformation	40
3.6	Poincaré's Time Transformation	41
3.7	Hamilton–Jacobi Equation	42
4	**Variation of Parameters**	**43**
4.1	Lagrange's Formulation	43
4.2	Poisson's Equations	45
4.3	Herrick's Perturbative Variation	46
4.4	Perturbations of Orbital Elements	50

5 Numerical Integration 55

5.1 Difference Methods 55
5.2 Runge–Kutta and Power Series Methods 62
5.3 Extrapolation Methods and Substep Algorithms 72
5.4 High-Order Leapfrogs 78

6 Symplectic Integration 81

6.1 Basic Theory 83
6.2 Composite Symplectic Methods 87
6.3 Non-Canonical Forces 104
6.4 Alternative Formulations 107
6.5 Near-Integrable Systems 112
6.6 Time Transformations 122
6.7 The Logarithmic Hamiltonian 131
6.8 Functional Time Transformation 135

7 KS-Regularization 143

7.1 Extended Phase Space and Time Transformations 143
7.2 KS-Transformation 144
7.3 Binaries as Perturbed Harmonic Oscillators 150
7.4 Multiparticle Regularization 153
7.5 Slow-Down 170

8 Algorithmic Regularization 173

8.1 The Basic Leapfrog Algorithms 173
8.2 Generalized Composite Leapfrog 175
8.3 Algorithmic Regularization 177
8.4 Logarithmic Hamiltonian Derivations 178
8.5 Logarithmic Leapfrog 180
8.6 Time-Transformed Leapfrog 185
8.7 Velocity-Dependent Perturbations 194
8.8 Special Cases 198
8.9 Tidal Friction and Spin of Stars 203
8.10 An Efficient Three-Body Algorithm 204
8.11 Formulation with Relative Coordinates 206
8.12 Algorithms 207

9 Motion in the Field of a Black Hole 211

9.1 Post-Newtonian Terms 211

10 Artificial Satellite Orbits **221**

10.1 Logarithmic Hamiltonian 221
10.2 The Exact Two-Body Leapfrog 223
10.3 Perturbed Two-Body Motion 224
10.4 Application to Artificial Satellites 228
10.5 Methods and Functions 234

References 237
Index 243

Preface

This book contains an exposition of the most important subjects in celestial mechanics that I have been working on. The book also includes some traditional material, which is necessary for anyone who wants to learn/study the gravitational few-body problem. Some basic subjects are repeated in different chapters/sections to avoid the need to read the entire book if one is interested only in some special problem.

Since 1972 I have been interested in the regularization of equations of motion for the gravitational few-body problem. The starting point was the excellent book by Stiefel and Scheifele (1971). In the days when I started my PhD Thesis work on the four-body problem I learned how to use the so-called KS-transformation (Kustaanheimo and Stiefel, 1965) on the few-body problem. My first achievement was the programming of the complete regularization of the N-body problem due to Heggie (1974). Later I developed a different method, the chain method, in collaboration with Sverre Aarseth during my first extended visit to Cambridge, UK (Mikkola and Aarseth, 1990).

More than a decade later I invented, completely accidentally, a new regularization method, which is called algorithmic regularization (AR). After completing the idea with Kiyotaka Tanikawa, the method was published in 1999 (Mikkola and Tanikawa, 1999a) but is not yet a textbook subject and is therefore a quite important part of this book.

My first astronomy professor Liisi Oterma (1915–2001) taught me how to compute orbits of asteroids from observations. This was the starting point of my interest in celestial mechanics, especially in the numerical methods.

Professor Mauri Valtonen suggested the four-body problem as the topic of my PhD work. That led me to learn the practical use of the KS-transformation.

Soon after that I visited Sverre Aarseth in Cambridge. This visit led to a long-time collaboration with Aarseth and, after some time, led to the invention of the chain method and later to a generalization of the AR method, i.e. the time-transformed leapfrog (TTL).

Professor Kiyotaka Tanikawa was involved with the invention of the AR and in addition we were working on the mathematics of the three-body problem, especially in the one- and two-dimensional cases which are often non-physical but interesting mathematically.

Work with David Merritt led to the writing of the Algorithmic Regularization Chain code (ARCHAIN) and to the invention of methods that made it possible to include velocity-dependent perturbation into AR, in which it was originally excluded.

For the introduction to satellite dynamics and collaboration I thank Phil Palmer.

I also would like to thank Kimmo Innanen (1937–2011) who encouraged me to study the dynamics of asteroids. This led me to familiarize myself with the symplectic integrators, especially with the Wisdom–Holman method (Wisdom and Holman, 1991). An additional consequence of that was my later invention of a partial regularization of the Wisdom–Holman method with the help of a time transformation.

I am grateful to all the above-mentioned teachers and colleagues, Kimmo Innanen, Sverre Aarseth, David Merritt, Kiyotaka Tanikawa, Phil Palmer, Mauri Valtonen and Liisi Oterma for their support and interest in my work. Particularly I thank Kimmo Innanen, Sverre Aarseth, Bo Reipurth and Kiyotaka Tanikawa for their help with the writing of this book.

Seppo Mikkola

Introduction

In Chapter 1 the basics of various few-body problems are discussed. These include planetary systems and small stellar clusters. The two-body problem is an important part for understanding few-body dynamics, and hence Chapter 2 discusses this subject. Additionally Chapter 2 covers the mathematics needed to compute the trajectory, orbital elements and some special functions which are convenient to evaluate the movement in elliptical, parabolic and hyperbolic trajectories. The basic classical mathematics of gravitational systems is discussed in Chapter 3. The analytical approximations, especially the 'variation of parameters' methods and corresponding differential equations, are the subject of Chapter 4. In Chapter 5 many classical numerical methods are discussed starting from difference methods which are accurate in systems that behave similarly over the studied time interval. In few-body stellar dynamics the systems often change significantly and a constant stepsize cannot be used for long times. The solution has been that of using extrapolation methods in which one uses first simple time-symmetrical integrators with shorter and shorter stepsizes to proceed over a given interval and then one uses polynomial or rational function extrapolation to estimate what the result would be at zero stepsize. In the Solar System, orbits remain mostly similar and therefore the method of symplectic integration (Chapter 6) is found to be a most efficient way for computation. With any computational method it is often necessary to employ coordinate and/or time transformations. There are three ways to proceed here: in Chapter 7 using the KS-transformation (KS) with a time transformation; in Chapter 8, algorithmic regularizations (AR) using the logarithmic Hamiltonian (logH) or the time-transformed leapfrog (TTL), which both are essentially based on regularizing time transformations which give exact trajectories for the two-body problem and generally regular results suitable for extrapolation algorithms. In Chapter 9 the dynamics near black holes is discussed in terms of the post-Newtonian formalism. Also the evolution of black hole spin is handled here. Finally in Chapter 10 a computational method for calculating the motion of artificial satellites in the complicated gravitational field of the Earth is presented. This can be applied in any field given in the form of an expansion in terms of the spherical functions.

Codes and additional information for this book are available at www.astro.utu.fi/mikkola/.

Chapter 1
The Problems

1.1 Introductory Presentation of the Problems

1.1.1 Units

In this book one choice of units to be especially mentioned is that $G = 1$, i.e. the units are such that the gravitational constant is 1. One convenient set of units is:

Gravitational constant $G = 1$.
Unit of mass is that of the Sun, i.e. $M_{\odot} = 1$.
Unit of distance = 1 AU.
This choice yields for the unit of time: a year $= 2\pi$.

An additional convenient coincidence is that in this system the speed of light is $c \approx 10^4$, which is easy to remember.

1.1.2 Problems of Celestial Mechanics and Few-Body Dynamics

The laws of motion of celestial bodies are differential equations. Normally these cannot be solved analytically. The only real-world example of a solvable case is the two-body motion: in a binary, far away from all other stars, the relative motion of the two stars is elliptical following Kepler's laws. Approximately this is true also in planetary systems, such as our Solar System. If the mutual interactions of the planets are taken into account, analytical solution of the problem is not possible any more.

Therefore, to study the behaviour of more complex systems than the two-body motion, we need numerical methods. These are discrete approximations for integration of the differential equations describing the system. The most common way to construct such approximations has been the use of polynomials, replacing the exact expressions of derivatives either by interpolation polynomials (which are easy to integrate), or by Taylor polynomials. In recent times other types of methods, composite and symplectic, have gained popularity. These methods have the advantage that they can more easily be adapted to the special properties, such as symmetries, of a system and in fact significantly more reliable results can be obtained in long-term computations.

Another direction of research has focused on the regularization of equations of motion. These are techniques that remove, or make milder, singularities of equations of motion. Such a singularity is a collision of bodies. In practice, however, exact collision is not required but a close approach is enough to cause serious trouble for most integration methods.

1.1.3 The Two-Body Problem

The two-body problem can be solved analytically, but only as long as it is unperturbed. In more complicated systems, exact analytical methods are not available and therefore one needs numerical methods.

Figure 1.1 describes how much the speeds of bodies in an eccentric orbit vary depending on the distance. The high speed when bodies are close to each other is what affects the accuracy of numerical computations. Very eccentric two-body orbits are not easily integrated numerically. Therefore the first step in developing reliable methods is to find a procedure that is regular for the two-body problem independent of eccentricity. It is necessary that the method can be generalized such that it works also for the perturbed two-body problem. Such a regularized method was first invented by Paul Kustaanheimo (and published in Kustaanheimo and Stiefel (1965) (KS)). A decade later the KS method was generalized by Aarseth and Zare (1974) and Heggie (1974) for systems with more than two bodies. A different regular method was later published by Mikkola and Tanikawa (1999a) and Preto and Tremaine (1999).

The equations of motion for the relative vector $\mathbf{r}$ between two masses m_1 and m_2 is

$$\ddot{\mathbf{r}} = -(m_1 + m_2)\mathbf{r}/r^3. \tag{1.1}$$

This looks very simple, but its full analytical solution is quite complicated. Therefore this problem is discussed quite deeply in Chapter 2.

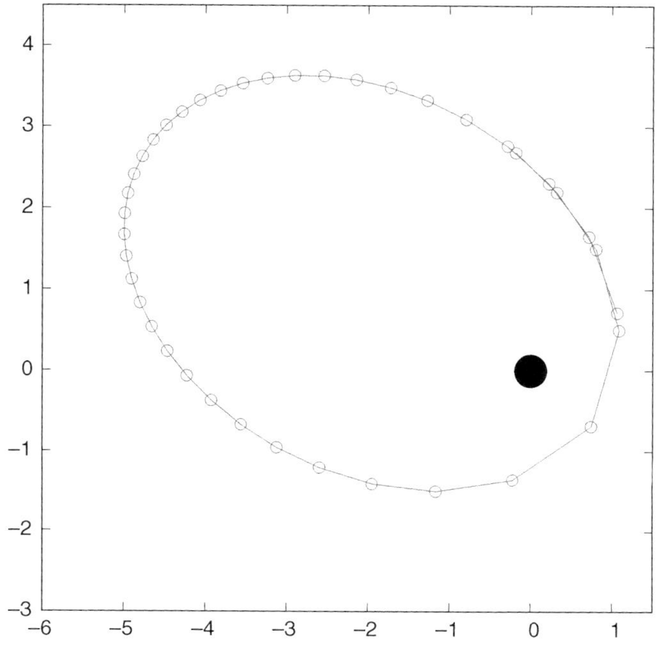

Fig. 1.1. A two-body orbit. The open circles show the body positions at equidistant times. It is clear that the speed is much higher near the pericentre.

1.1.4 The N-Body Problem

The N-body problem, with body masses m_k, coordinates $\mathbf{r}_k$, velocities $\mathbf{v}_k$ and momenta $\mathbf{p}_k = m_k \mathbf{v}_k$, can be defined by its Hamiltonian

$$H = \sum_k \mathbf{p}_k^2/(2m_k) - \sum_{i<j} m_i m_j / r_{ij}, \tag{1.2}$$

where $r_{ij} = |\mathbf{r}_j - \mathbf{r}_i|$ are the relative distances. The equations of motion are

$$\dot{\mathbf{r}}_k = \frac{\partial H}{\partial \mathbf{p}_k} = \frac{\mathbf{p}_k}{m_k} = \mathbf{v}_k, \tag{1.3}$$

$$\dot{\mathbf{p}}_k = -\frac{\partial H}{\partial \mathbf{r}_k} = \sum_{i \neq k} m_i m_k \frac{\mathbf{r}_i - \mathbf{r}_k}{r_{ki}^3}, \tag{1.4}$$

or in the form of second-order equation

$$\ddot{\mathbf{r}}_k = \sum_{i \neq k} m_i \frac{\mathbf{r}_i - \mathbf{r}_k}{r_{ki}^3}. \tag{1.5}$$

In Figure 1.2 is illustrated a complicated five-body orbit. The solution method for such an orbit is not easy but can be done with efficient methods that are discussed in this book.

As is well known the integrals of motion in an N-body problem are (in obvious notation):

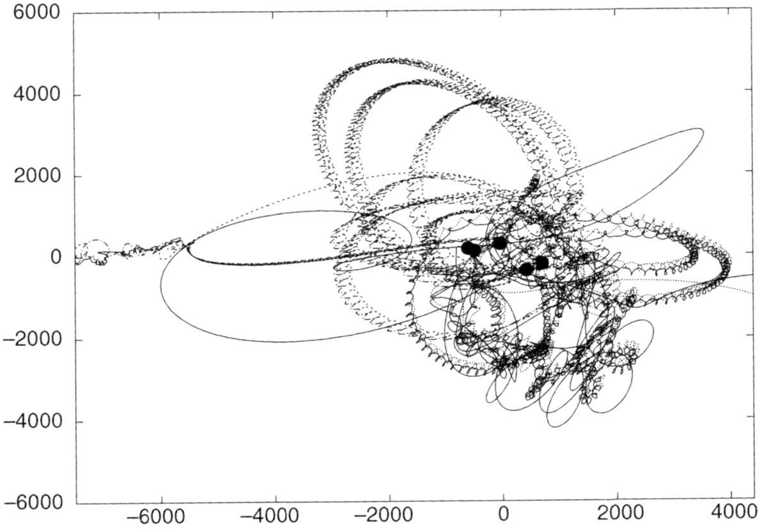

Fig. 1.2. A three-dimensional five-body orbit. One can see that there are situations of close approaches which probably could not be integrated without regularization. The starting points of the five bodies are shown as big dots. The final result of the interaction is one binary and three single ones.

The centre of mass

$$\mathbf{r}_{cm} = \sum_{k=1}^{N} m_k \mathbf{r}_k / m_{tot} = \mathbf{r}_{cm}(0) + \mathbf{v}_{cm}(0)t. \tag{1.6}$$

Energy integral

$$H = \sum_k \mathbf{p}_k^2/(2m_k) - \sum_{i<j} m_i m_j / r_{ij} = \text{constant}. \tag{1.7}$$

Angular momentum

$$\mathbf{L} = \sum_{k=1}^{N} m_k \mathbf{r}_k \times \mathbf{v}_k = \text{constant}. \tag{1.8}$$

In general all integrals of motion are related to (in fact consequences of) symmetries of the system. The linear motion of the centre of mass is due to the translation symmetry of the system, i.e. the origin of coordinates does not change the Hamiltonian or equations of motion. Similarly the conservation of angular momentum exists due to rotational symmetry of the system. Finally the energy is constant because of the time-reversal symmetry of the system: a change of sign of time does not affect the equations describing the system.

These symmetries are important in the numerical methods too: If the discrete method, with which we approximate the solution for the equations of motion, has the same symmetries as the original system, we can often expect better conservation of the constants of motion. Many reasonable methods conserve the centre-of-mass velocity and also the angular momentum exactly. The energy is more difficult to conserve and classical methods (polynomial-based difference and Taylor series) typically produce a secular energy, which means the existence of spurious friction in the method. However, if this effect is small enough, such methods can be satisfactory.

In the past it was customary to avoid time reversibility of the method. The argument was that then one loses the possibility of using a time-reversal test for the accuracy of the method. Nowadays, contrary to past beliefs, it has been recognized that in fact the time reversibility of an integrator is desirable. Due to the connection with the conservation of energy and time reversibility such methods perform well in this respect. The quantity conserved by the discrete integrator is usually not exactly the energy of the considered system, but some other 'nearby' expression which, however, tends to the energy with decreasing stepsize of the integrator.

1.1.5 Planetary Dynamics

An important class of orbits are the planetary ones which are illustrated in Figure 1.3 They are known to be elliptical, but in fact the

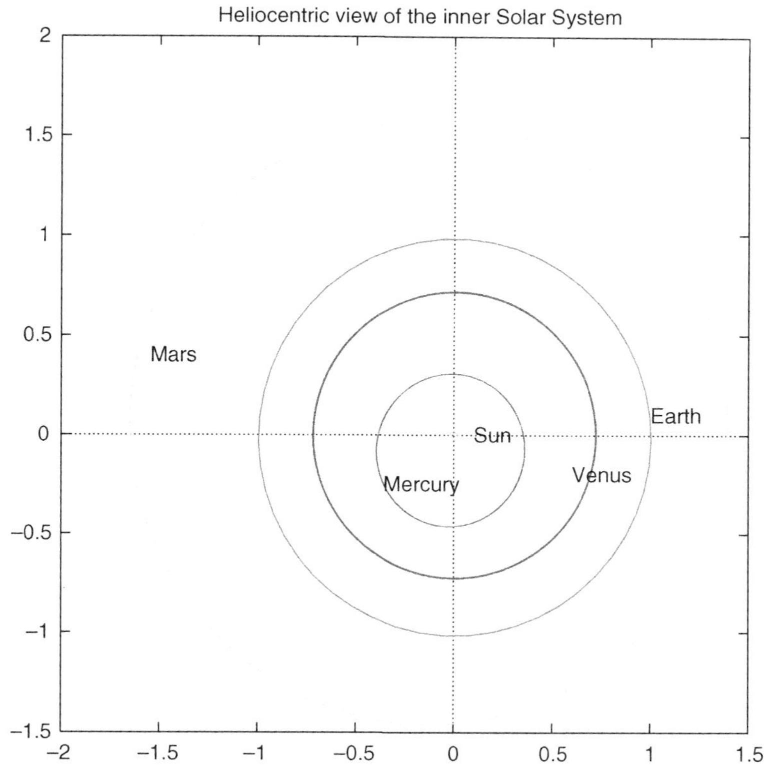

Fig. 1.3. The inner Solar System. The fairly high eccentricity of the planet Mars made it possible for Johannes Kepler to realize that the planetary orbits are elliptic.

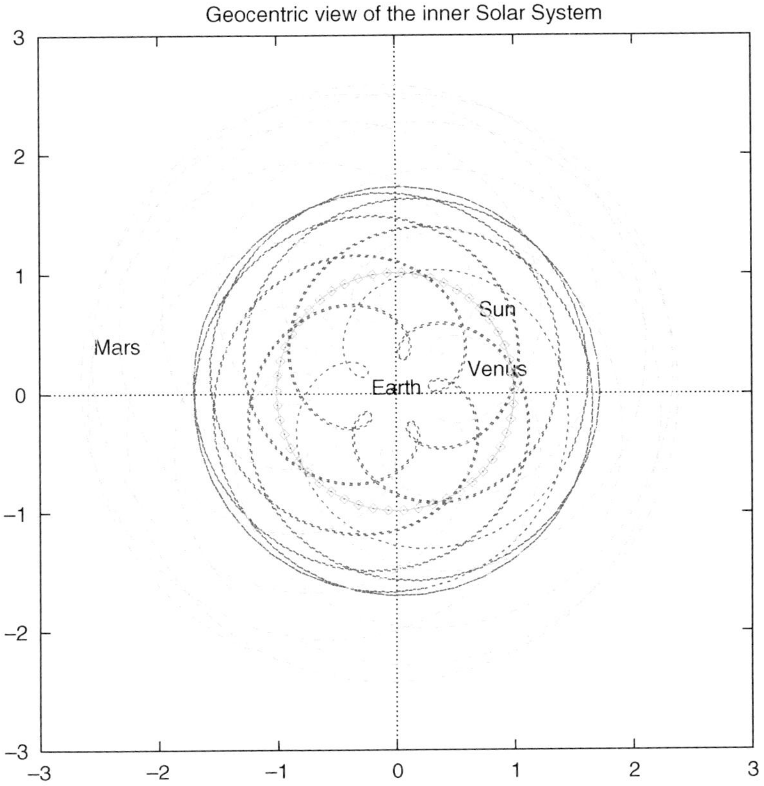

Fig. 1.4. The inner Solar System as seen in the Earth-centred system. This is what one can see from the Earth (actually only the directions of planetary positions).

ellipses keep changing all the time due to the perturbations due to other planets. A class of special numerical methods for planetary orbits is called symplectic and is discussed in Chapter 6.

In the times before Isaac Newton published his 'Philosophiae Naturalis Principia Mathematica' the motions of the planets were considered to happen around the Earth. This is geometrically completely correct but more complicated (see Figure 1.4) than the later models of elliptical motion around the Sun. There were also suggestions that planets move around the Sun, but the Sun moves around the Earth (theory of Tycho Brahe (1546–1601)).

A topic in N-body dynamics is the problem of computing planetary motions. The basic equations are the same as those in the N-body problem. However, there is one dominating body (the Sun) and all other bodies are relatively small. Therefore it is often convenient to consider the orbits as perturbed two-body trajectories. Quite often the equations of motion are written in heliocentric coordinates

$$\ddot{\mathbf{r}}_k = -(M_\odot + m_k)\mathbf{r}_k/r_k^3 - \sum_{i \neq k} m_i((\mathbf{r}_k - \mathbf{r}_i)/r_{ki}^3 - \mathbf{r}_i/r_i^3), \qquad (1.9)$$

where $\mathbf{r}_k$ are the position vectors of the planets in a coordinate system with the Sun at the origin. Note that for a massless test particle (or for a practically massless body, such as an asteroid or a comet) the mass $m_k = 0$, but otherwise the equation is the same.

Another set of commonly used coordinates is the Jacobian coordinate system where each planet is referred to the centre of mass of the 'lower' planets plus the Sun. In these variables the two-body orbits are typically even better approximations to the real motion than in the heliocentric coordinates. The transformation is

$$\mathbf{r}_i^J = \mathbf{r}_i - \mathbf{X}_i, \quad \mathbf{X}_i = \frac{\sum_{k=0}^{i-1} m_k \mathbf{r}_k}{\sum_{k=0}^{i-1} m_k}, \tag{1.10}$$

and the Hamiltonian

$$H_J = \frac{\mathbf{p}_{cm}^2}{2M_{cm}} + \sum_{i=0}^{n} \frac{\tilde{\mathbf{p}}_i^2}{2\mu_i} - \sum_{i<j} m_i m_j / r_{ij}, \tag{1.11}$$

where μ_i is the reduced mass

$$\mu_i = m_i M_{i-1}/M_i, \quad M_j = \sum_{k=0}^{j} m_k. \tag{1.12}$$

The centre-of-mass-related quantities $\mathbf{p}_{cm}$ and M_{cm} may be given by

$$\mathbf{p}_{cm} = \sum \mathbf{p}_i, \quad M_{cm} = M_n. \tag{1.13}$$

The relative distances r_{ij} are expressed in terms of the Jacobian coordinates $\mathbf{r}_i^J$ and the Jacobian momenta are

$$\tilde{\mathbf{p}}_i = \mu_i \dot{\mathbf{r}}^J. \tag{1.14}$$

1.1.6 Satellite Dynamics

A practically important problem is that of computing the motions of satellites, natural and artificial. Although the equations of motion contain N-body terms (due to attractions from all the bodies in the system), the problem is often dominated by the perturbation due to flattening of the planet. In the case of artificial Earth satellites this is the so-called J_2-problem: The Hamiltonian of the system is

$$H = \frac{\mathbf{p}^2}{2} - \frac{M}{r} + \frac{MJ_2R^2}{2r^3}\left(1 - 3\left(\frac{z}{r}\right)^2\right) + \text{higher-order terms}, \tag{1.15}$$

where M is the Earth's mass, R is Earth's radius and J_2 is a dimensionless constant, which determines the size of the perturbation

($J_2 \approx -10^{-3}$). There is also a huge number of much smaller, but non-negligible additional terms (of order $O(J_2^2)$) in the Earth's potential (and the Earth rotates!). Also atmospheric friction and effects due to the Moon, Sun and planets should be considered.

We see that essentially this is a problem of perturbed two-body motion, for which there exist specific methods that take into account the fact that most of the force field is integrable. Many other problems, including the planetary motion and binary motion in stellar systems, may also be considered as perturbed two-body problems and similar methods could be tried for the computation of motion in such systems.

1.1.7 Chaos in N-Body Systems

Most N-body systems, already for $N = 3$, are very sensitive to any tiny change of initial conditions or external perturbations. They are *chaotic*. In such a situation the trajectories of two almost the same systems separate exponentially with time. Thus the magnitude of the separation (δq) behaves like

$$\delta q \sim c \exp(\lambda t),$$

where c is some constant, t is time and λ is the variational exponent which tells the rate of chaos. The inverse $1/\lambda$ is then the so-called e-folding time of the difference. That is an important measure of the chaoticity of the system; if it is small the system is very chaotic. In Sections 5.3.7 and 6.2.5 there are explanations on how to determine the chaoticity. Often N-body systems are so chaotic that accurate numerical computation of the trajectories is in practice not possible. It is anyway widely believed that statistically the simulation results are acceptable. One argument for this is that in the real physical systems, star clusters, there is nothing inexplicable although those systems are chaotic and the forces in them are complicated, not only point-mass forces, but many small tidal effects and external perturbations due to other masses.

1.2 Historical (a Point of View of the Author)

Some major stepping stones and events in the development of the numerical tools for the investigation of few-body dynamics, especially strong interactions, are listed below.

1. von Hoerner (1960, 1963) discovered that the formation of binaries in few-body systems is a rule rather than an exception. However, von Hoerner was unable to study the process in detail due to computational difficulties: accurate numerical integration of the motion of (eccentric) binaries was too difficult with the numerical tools available at the time.

2. Kustaanheimo and Stiefel (1965) published their regularization method (generalization of the celebrated two-dimensional Levi-Civita transformation), for the (perturbed) two-body motion, which made possible accurate calculations of close encounters and binary motions in clusters.

3. Aarseth and Zare (1974) published a three-particle regularization method. The two shortest distances are regularized with the Kustaanheimo–Stiefel (KS) method and the arrangement is appropriately updated when necessary. Also Zare (1974) generalized this method by regularizing an arbitrary number of bodies with respect to one dominating body.

4. Heggie (1974) realized that it is possible to use an extended phase space and regularize all the interparticle vectors in the system simultaneously, thus achieving a global regularization.

5. Mikkola (1985) was the first to write a concise regularized algorithm for N bodies utilizing Heggie's global method.

6. Mikkola and Aarseth (1990, 1993) developed what is known as the *chain method*. This method generalizes the idea of Aarseth and Zare (1974). Forming a chain of interparticle vectors and applying the KS-transformation to each of these, the authors obtained a practical regularized N-body method without extra degrees of freedom.

7. Wisdom and Holman (1991) introduced a new method (the WH method), the generalized leapfrog, which is a symplectic map. Contrary to earlier symplectic integrators this is both faster and, in long integrations, a reliable way for studying planetary-type systems.

8. Mikkola and Aarseth (1996) further upgraded their method by introducing the idea of *slowing down* tight binaries. This method preserves the secular perturbations, but loses the phase of the binary. Due to the chaos in many-body systems, the phase of a binary should not be statistically important, however.

9. Mikkola (1997) applied time transformation to the WH method, thus producing a symplectic integrator which is stable also for eccentric orbits.

10. Mikkola and Aarseth (1998) successfully applied the Stumpff functions in integration of binaries. This gives an advantage especially in the case of well-isolated binaries.

11. Mikkola and Tanikawa (1999a) and Preto and Tremaine (1999) introduced a modified, time-transformed (by using logarithmic Hamiltonian) (logH) leapfrog which accurately follows two-body orbits. This can also be applied more generally (Mikkola and Tanikawa, 1999b), with extrapolation methods, to produce regular few-body algorithms without KS-regularization. Mikkola and Tanikawa (1999a,b) suggest the name *algorithmic regularization* (AR) for this.

12. Mikkola and Aarseth (2002) introduced the time-transformed leapfrog (TTL).

13. Mikkola and Merritt (2006) applied a generalized midpoint method to include velocity-dependent perturbations into an algorithmic regularization code.

14. Hellström and Mikkola (2009) published the idea of auxiliary velocity which makes it possible to construct a leapfrog with velocity-dependent perturbations and thus use algorithmic regularization even in that case. This method is more elegant and efficient than the earlier generalized midpoint.

15. Aarseth (2010) published a book that contains also many of the formulae that are here used in Section 7.4. Those results are also in the many earlier papers by Mikkola and Aarseth.

In the N-body problems with large N, one usually uses the so-called Hermite methods (Makino and Aarseth, 1992) (it is outside

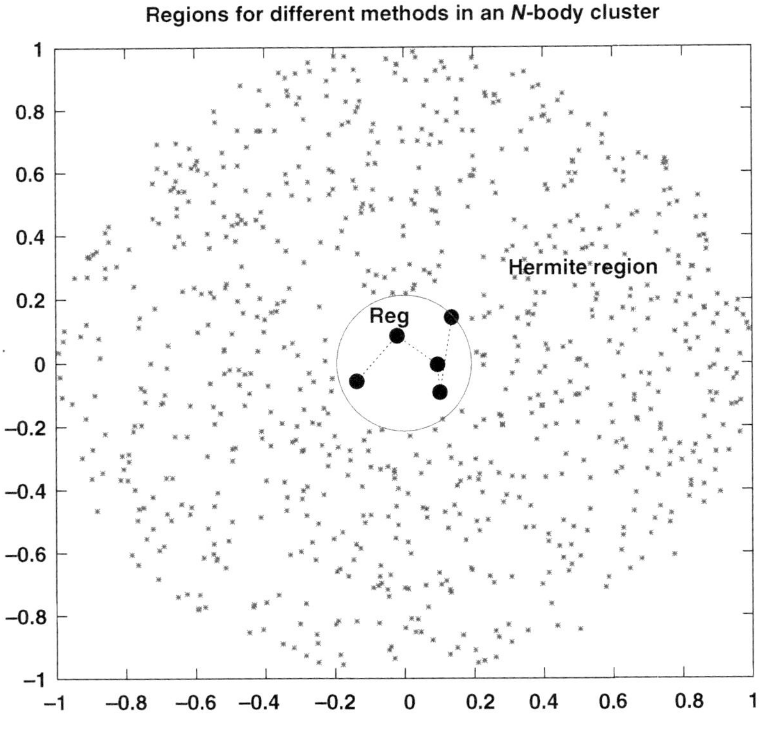

Fig. 1.5. Regions in an N-body cluster in which different integration methods are often used to make the results reliable.

the scope of this book). In Figure 1.5 the different regions for the Hermite and regularized methods (which are one of the subjects of this book) are useful. Thus the regular methods (KS and AR) discussed in later chapters (Chapters 7 and 8) are good in the most energetic regions of stellar clusters: in the innermost ones.

Chapter 2
Two-Body Motion

The two-body problem is the only N-body problem which has an analytic solution and many developments in celestial mechanics have their basis in the solution of this problem. Thus it is of utmost importance to know the details of this special case. This chapter lists many of the important formulae related to two-body motion. More information can be found, for example, in Stumpff (1959) and in Danby (1992).

According to Newton's law two bodies of masses m_1 and m_2 attract each other with the force Gm_1m_2/r^2, where r is the distance between their mass centres and G is the gravitational constant. For simplicity, units in which $G = 1$ are adopted. Thus, the equations of motion read

$$m_1\ddot{\mathbf{r}}_1 = -m_1m_2\frac{\mathbf{r}_1 - \mathbf{r}_2}{r^3}, \quad m_2\ddot{\mathbf{r}}_2 = -m_1m_2\frac{\mathbf{r}_2 - \mathbf{r}_1}{r^3}, \tag{2.1}$$

where $r = |\mathbf{r}_1 - \mathbf{r}_2|$. From these, one easily derives the results

$$\ddot{\mathbf{r}}_{cm} = \mathbf{0}, \quad \ddot{\mathbf{r}} = -\mu\frac{\mathbf{r}}{r^3}, \tag{2.2}$$

where $\mu = m_1 + m_2$ is the total mass, $\mathbf{r}_{cm} = (m_1\mathbf{r}_1 + m_2\mathbf{r}_2)/\mu$ is the centre of mass and $\mathbf{r} = \mathbf{r}_2 - \mathbf{r}_1$ is the relative position vector. Thus the centre of mass moves with constant velocity, while the relative motion is to be solved from the above differential equation.

2.1 Solution of Equations of Motion

2.1.1 Form of the Orbit

Since the relative motion of two bodies satisfies the equation

$$\ddot{\mathbf{r}} = -\mu\mathbf{r}/r^3, \tag{2.3}$$

the derivative of the angular momentum per unit of mass $\mathbf{h} = \mathbf{r} \times \mathbf{v}$ (where $\mathbf{v} = \dot{\mathbf{r}}$ is the velocity) vanishes, because

$$\dot{\mathbf{h}} = \underbrace{\mathbf{v} \times \mathbf{v}}_{=0} + \mathbf{r} \times \ddot{\mathbf{r}} = -\mu \underbrace{\mathbf{r} \times \mathbf{r}}_{=0} / r^3 = 0. \tag{2.4}$$

Thus

$$\mathbf{h} = \mathbf{r} \times \mathbf{v} = \text{constant}. \tag{2.5}$$

The scalar product $\mathbf{r} \cdot \mathbf{h} = 0$, so that the motion is restricted to a plane perpendicular to $\mathbf{h}$. The conservation of the Kepler energy $\frac{1}{2}\mathbf{v}^2 - \mu/r$ may be demonstrated by differentiating with respect to the time

$$\frac{d}{dt}\left(\frac{1}{2}\mathbf{v}^2 - \frac{\mu}{r}\right) = \mathbf{v} \cdot \ddot{\mathbf{r}} + \mu\frac{\mathbf{r}}{r^3} \cdot \mathbf{v} = 0, \tag{2.6}$$

which is identically zero because $\ddot{\mathbf{r}} = -\mu\mathbf{r}/r^3$. Consider the cross product $\ddot{\mathbf{r}} \times \mathbf{h}$, which can be expanded as

$$\ddot{\mathbf{r}} \times \mathbf{h} = -\frac{\mu}{r^3}\mathbf{r} \times (\mathbf{r} \times \mathbf{v}) = -\frac{\mu}{r^3}((\mathbf{r} \cdot \mathbf{v})\mathbf{r} - (\mathbf{r} \cdot \mathbf{r})\mathbf{v}), \tag{2.7}$$

so that

$$\ddot{\mathbf{r}} \times \mathbf{h} = \frac{\mu}{r^2}(r\mathbf{v} - \dot{r}\mathbf{r}) = \mu\frac{d}{dt}(\mathbf{r}/r). \tag{2.8}$$

Integration produces

$$\mathbf{v} \times \mathbf{h} = \mu\left(\frac{\mathbf{r}}{r} + \mathbf{e}\right), \tag{2.9}$$

where the *eccentricity vector*

$$\mathbf{e} = \frac{\mathbf{v} \times \mathbf{h}}{\mu} - \frac{\mathbf{r}}{r} \tag{2.10}$$

is a constant of integration. We find $\mathbf{v} \times \mathbf{h} \cdot \mathbf{r} = \mu(r + \mathbf{e} \cdot \mathbf{r})$, where the left side is h^2. Thus if v is the angle between the vectors $\mathbf{e}$ and $\mathbf{r}$ (not to be confused with the velocity), then we have $h^2 = \mu r(1 + e\cos(v))$. Abbreviating $h^2/\mu = p$, one may thus write

$$r = p/(1 + e\cos(v)), \tag{2.11}$$

which is the equation of a conic section in polar coordinates (e = eccentricity, p = semi-latus-rectum, semi-parameter). Thus the orbit is an *ellipse*, *parabola*, or *hyperbola*, depending on the values of the integration constants. From the above expression one reads easily the well-known relations $q = p/(1+e) = a(1-e)$, the minimum distance, and $a = p/(1-e^2)$, the semi-major axis. In general, the position vector $\mathbf{r}$ can be written as

$$\mathbf{r} = x_\omega \mathbf{P} + y_\omega \mathbf{Q}, \tag{2.12}$$

where $\mathbf{P}$ and $\mathbf{Q}$ are the directions of $\mathbf{r}$ when $v = 0°$, and when $v = 90°$, respectively, and where

$$x_\omega = r\cos(v), \quad y_\omega = r\sin(v). \tag{2.13}$$

2.1.2 Motion in Time

To obtain results that are valid for any type of orbit, a time transformation is introduced by writing $dt = rdX$. The first and second derivatives with respect to X (denoted by a prime) are: $\mathbf{r}' = r\mathbf{v}$ and $\mathbf{r}'' = r'\mathbf{v} + r\mathbf{v}' = \mathbf{r}\cdot\mathbf{v}\mathbf{v} + r^2\dot{\mathbf{v}} = (\mathbf{r}\cdot\mathbf{v})\mathbf{v} - \mu\mathbf{r}/r$. Adding the equation (2.10) for $\mathbf{e}$ into this, we obtain

$$\mathbf{r}'' + \beta\mathbf{r} = -\mu\mathbf{e}, \tag{2.14}$$

where $\beta = 2\mu/r - \mathbf{v}^2$ is a constant, in fact twice the negative of the Kepler energy. Thus the two-body motion has been reduced to a shifted harmonic oscillator when the new variable X is used as the independent variable. The dependence on time requires the solution of the time equation $t' = r$. To this end we calculate r''. First we have $r' = r\dot{r} = \mathbf{r}\cdot\mathbf{v}$ and thus $r'' = r\mathbf{v}^2 + r\mathbf{r}\cdot(-\mu\mathbf{r}/r^3) = r\mathbf{v}^2 - \mu$. The expression $\mathbf{v}^2 = 2\mu/r - \beta$ produces the simple result

$$r'' + \beta r = \mu. \tag{2.15}$$

If the equations (2.14) and (2.15) are differentiated repeatedly, one obtains

$$y^{(n+2)} + \beta y^{(n)} = 0, \tag{2.16}$$

where y is any of the coordinates or the distance r. From the recursion relation $y^{(n+2)} = -\beta y^{(n)}$ (for $n > 0$), the Taylor series solutions may be written

$$\mathbf{r} = \mathbf{r}_0 + \mathbf{r}_0' G_1 + \mathbf{r}_0'' G_2, \tag{2.17}$$

$$r = r_0 + r_0' G_1 + r_0'' G_2, \tag{2.18}$$

$$t = r_0 X + r_0' G_2 + r_0'' G_3, \tag{2.19}$$

where G_1 and G_2 are the power series

$$G_k = \frac{X^k}{k!} - \beta\frac{X^{k+2}}{(k+2)!} + \beta^2\frac{X^{k+4}}{(k+4)!} - \beta^3\frac{X^{k+6}}{(k+6)!} + \cdots, \quad (2.20)$$

which satisfy the simple differentiation rule $G_k' = G_{k-1}$. Using the relations $\mathbf{r}' = r\mathbf{v}$, $\mathbf{r}'' = -\beta r - \mu\mathbf{e}$, $\mu\mathbf{e} = \mathbf{v}^2\mathbf{r} - \mathbf{v}\cdot\mathbf{r}\mathbf{v} - \mu\mathbf{r}/r$ and the expression for time t, one can rewrite the solution for $\mathbf{r}$ in the form $\mathbf{r} = \mathbf{r}_0\,(1 - \mu G_2/r_0) + \mathbf{v}_0\,(t - \mu G_3)$ and the velocity may be given by $\mathbf{v} = \mathbf{r}'/r = \mathbf{r}_0(-\mu G_1/(r_0 r)) + \mathbf{v}_0(1 - \mu G_2/r)$. These relations are usually written

$$f = 1 - \frac{\mu G_2}{r_0}, \quad (2.21)$$

$$g = t - \mu G_3, \quad (2.22)$$

$$\dot{f} = -\frac{\mu G_1}{r_0 r}, \quad (2.23)$$

$$\dot{g} = 1 - \frac{\mu G_3}{r}, \quad (2.24)$$

$$\mathbf{r} = \mathbf{r}_0 f + \mathbf{v}_0 g, \quad (2.25)$$

$$\mathbf{v} = \mathbf{r}\dot{f} + \mathbf{v}_0\dot{g}. \quad (2.26)$$

These f and g functions will be discussed in detail in Section 2.4.1.

In addition to the above, the solutions for (2.14) and (2.15), in the case of positive β (elliptic motion), can be written

$$\mathbf{r} = \frac{-\mu\mathbf{e}}{\beta} + \left(\mathbf{r}_0 + \frac{\mu\mathbf{e}}{\beta}\right)\cos(\sqrt{\beta}X) + \frac{\mathbf{r}_0'}{\sqrt{\beta}}\sin(\sqrt{\beta}X), \quad (2.27)$$

$$r = \frac{\mu}{\beta} + \left(r_0 - \frac{\mu}{\beta}\right)\cos(\sqrt{\beta}X) + \frac{r_0'}{\sqrt{\beta}}\sin(\sqrt{\beta}X). \quad (2.28)$$

If the initial point is at pericentre, $r_0' = 0$ and $r_0 = q$. In this case $r = \mu/\beta + (q - \mu/\beta)\cos(\sqrt{\beta}X)$. Since the semi-major axis of an ellipse is the average of the minimum and maximum distances, $a = \mu/\beta$, which is usually written in the form of the so-called *vis-viva integral*

$$\frac{1}{a} = \frac{2}{r} - \frac{\mathbf{v}^2}{\mu}. \quad (2.29)$$

Since $q = a(1 - e)$ the distance r is given by

$$r = a(1 - e\cos(E)), \quad (2.30)$$

where $E = \sqrt{\beta}X$ is the eccentric anomaly. By integrating the equation $t' = r$, one obtains $t = \int r\, dX$ which results in *Kepler's equation*

$$E - e\sin(E) = M, \tag{2.31}$$

where $M = t - t_0/(a\sqrt{a/\mu})$ is the mean anomaly, and t_0 is the moment of pericentre passage. For the case of the corresponding hyperbola the hyperbolic Kepler's equation

$$e\sinh(F) - F = M = \frac{t - t_0}{A\sqrt{A/\mu}} \tag{2.32}$$

may be obtained from the elliptic equation by the transformation $F = iE$ and $A = -a$. The parabolic case must be handled using the constancy of angular momentum and (2.11). The result is *Barker's equation*

$$\frac{\sqrt{\mu}(t - t_0)}{q\sqrt{2q}} = \tan\left(\frac{v}{2}\right) + \frac{1}{3}\tan^3\left(\frac{v}{2}\right). \tag{2.33}$$

In addition to, and instead of, the true anomaly v, one can use the eccentric anomaly E for elliptic motion, and the hyperbolic anomaly F in a hyperbolic trajectory, to express the coordinates and velocities in one of the forms

$$x_\omega = r\cos(v) = a(\cos(E) - e) = A(e - \cosh(F)), \tag{2.34}$$

$$y_\omega = r\sin(v) = a\sqrt{1 - e^2}\sin(E) = A\sqrt{e^2 - 1}\sinh(F), \tag{2.35}$$

$$\mathbf{r} = x_\omega \mathbf{P} + y_\omega \mathbf{Q}, \tag{2.36}$$

$$\mathbf{v} = \dot{x}_\omega \mathbf{P} + \dot{y}_\omega \mathbf{Q}, \tag{2.37}$$

where the time derivatives of the coordinates are

$$\dot{x}_\omega = \frac{\partial x_\omega}{\partial v}, \quad \dot{v} = \frac{\partial x_\omega}{\partial E}, \quad \dot{E} = \frac{\partial x_\omega}{\partial F}\dot{F} \tag{2.38}$$

(and similar expressions for y_ω). The derivatives of v, E or F can be calculated from the corresponding Kepler's equation, e.g. $\dot{E} = 1/(r\sqrt{a/\mu})$. Finally, in an ellipse (Figure 2.1), within one period, the advancement of M is $\Delta M = \Delta E = 2\pi$. Thus the period P is given by

$$P = 2\pi a\sqrt{a/\mu}. \tag{2.39}$$

2.2 Orbital Elements

The complete specification of an orbit requires six (independent) constants of motion. These are called the orbital elements. Often

Fig. 2.1. Geometry of ellipse.

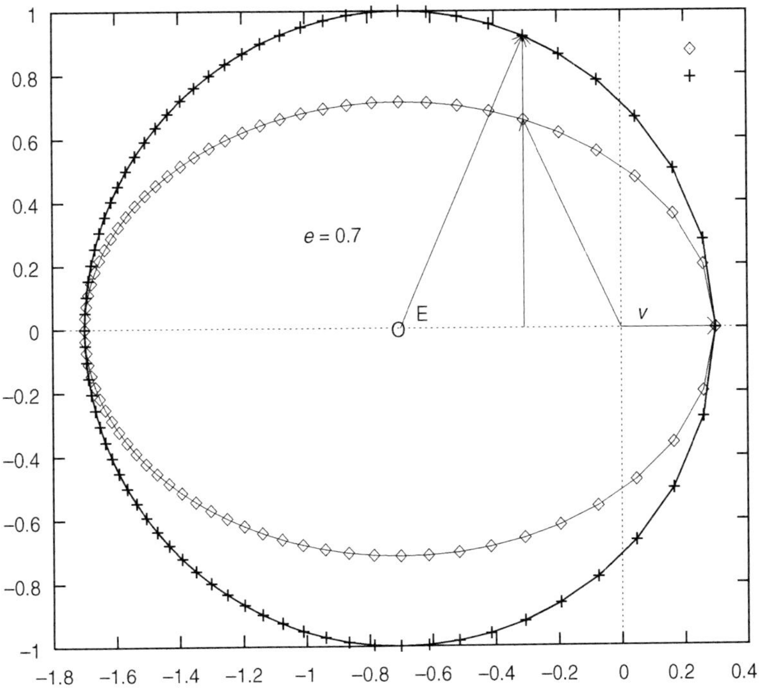

the so-called *Keplerian elements* are adopted, although any set of constants uniquely defining the orbit and time of motion would do as well.

The Keplerian or *elliptic elements* are:

$$a,\ e,\ i,\ \omega,\ \Omega \text{ and } t_0, \tag{2.40}$$

a = semi-major axis,
e = eccentricity,
i = inclination $0 < i < 180^\circ$,
ω = longitude of perihelion $0 < \omega < 360^\circ$,
Ω = longitude of the ascending node $0 < \Omega < 360^\circ$,
t_0 = time of pericentre passage (or instead $M_0 = M(t_0)$).

Other abbreviations often used are:

$\varpi = \omega + \Omega$,
$\lambda = M + \varpi$ is the mean longitude,
$n = \dot{M} = 1/(a\sqrt{a/\mu})$ is the mean motion.

The vectors **P** and **Q** and $\mathbf{W} = \mathbf{P} \times \mathbf{Q}$ (= normal of the orbital plane) are related to the angular elements (see Figure 2.2) via

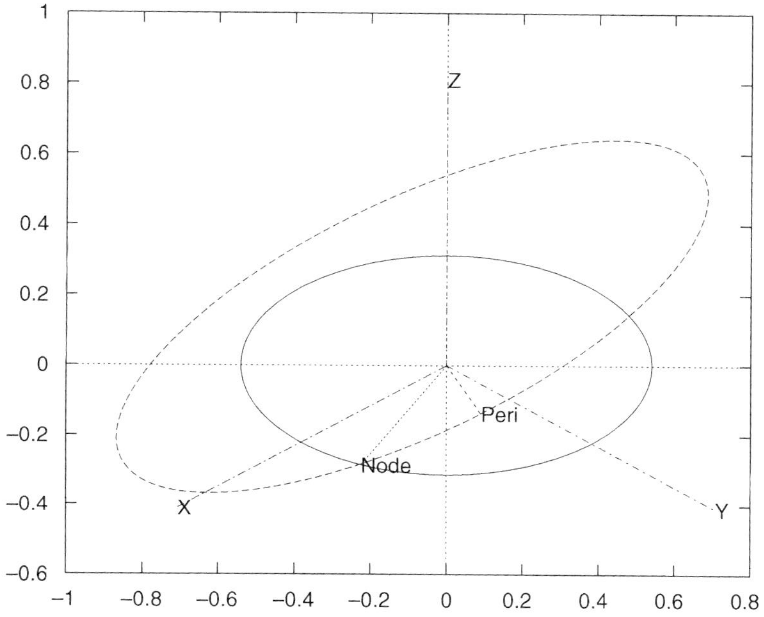

Fig. 2.2. Definition of the angular elements.

$$
\begin{aligned}
P_x &= +\cos(\omega)\cos(\Omega) - \sin(\omega)\sin(\Omega)\cos(i),\\
P_y &= +\cos(\omega)\sin(\Omega) + \sin(\omega)\cos(\Omega)\cos(i),\\
P_z &= +\sin(\omega)\sin(i),\\
\\
Q_x &= -\sin(\omega)\cos(\Omega) - \cos(\omega)\sin(\Omega)\cos(i),\\
Q_y &= -\sin(\omega)\sin(\Omega) + \cos(\omega)\cos(\Omega)\cos(i),\\
Q_z &= +\cos(\omega)\sin(i),\\
\\
W_x &= +\sin(\Omega)\sin(i),\\
W_y &= -\cos(\Omega)\sin(i),\\
W_z &= +\cos(i),
\end{aligned}
\tag{2.41}
$$

which relations follow from elementary geometrical considerations.

The direction vectors $\mathbf{P}$ and $\mathbf{W}$ are related to the eccentricity vector $\mathbf{e}$ and specific angular momentum $\mathbf{h}$ by the simple expressions $\mathbf{P} = \mathbf{e}/e$ and $\mathbf{W} = \mathbf{h}/h$. (Note: The eccentricity vector is called by many names, e.g. the Laplace–Runge–Lenz vector.) With this and the relation $\mathbf{Q} = \mathbf{P} \times \mathbf{W}$, one can compute the above vectors from the position and velocity. For the orbital elements we can thus derive formulae which are listed below:

Writing $\mathbf{w} = \mathbf{v}/\sqrt{\mu}$, then we have

$$\alpha = 2/r - \mathbf{w}^2, \tag{2.42}$$
$$a = 1/\alpha, \tag{2.43}$$
$$\eta = \mathbf{r} \cdot \mathbf{w}, \tag{2.44}$$
$$\zeta = 1 - \alpha r, \tag{2.45}$$
$$\mathbf{D} = \mathbf{r} \times \mathbf{w}, \quad D = |\mathbf{D}|, \tag{2.46}$$
$$\mathbf{e} = \mathbf{w} \times \mathbf{D} - \mathbf{r}/r, \tag{2.47}$$
$$B = \sqrt{D_x^2 + D_y^2}, \tag{2.48}$$
$$e = |\mathbf{e}| = \sqrt{e_x^2 + e_y^2 + e_z^2} = \sqrt{\zeta^2 + \alpha\eta^2}, \tag{2.49}$$
$$M_0 = \operatorname{atan}_2(\eta\sqrt{\alpha}, \zeta) - \eta\sqrt{\alpha}, \tag{2.50}$$
$$i = \operatorname{atan}_2(B, D_z), \tag{2.51}$$
$$\Omega = \operatorname{atan}_2(D_x, -D_y), \tag{2.52}$$
$$\omega = \operatorname{atan}_2(e_z D, e_y D_x - e_x D_y). \tag{2.53}$$

In an ellipse, the position and velocity at any given time may be evaluated by

$$\mathbf{P} = \mathbf{e}/e, \quad \mathbf{Q} = \mathbf{D} \times \mathbf{P}/D, \tag{2.54}$$
$$\mathbf{a} = a\mathbf{P}, \quad \mathbf{b} = a\sqrt{1 - e^2}\mathbf{Q}, \tag{2.55}$$
$$\mathbf{r} = \mathbf{a}(\cos(E) - e) + \mathbf{b}\sin(E), \quad E - e\sin(E) = M, \tag{2.56}$$
$$r = a(1 - e\cos(E)), \tag{2.57}$$
$$\mathbf{w} = (-\mathbf{a}\sin(E) + \mathbf{b}\cos(E))/(r\sqrt{a}), \tag{2.58}$$
$$\mathbf{v} = \sqrt{\mu}\mathbf{w}. \tag{2.59}$$

For a hyperbola, the calculation of the elements uses

$$A = 1/|\alpha|, \tag{2.60}$$
$$M_0 = \eta\sqrt{|\alpha|} - \operatorname{atanh}(\eta\sqrt{|\alpha|}/\zeta). \tag{2.61}$$

The other formulae are the same as above. The position–velocity computation can be done by

$$\mathbf{a} = A\mathbf{P}, \tag{2.62}$$
$$\mathbf{b} = A\sqrt{e^2 - 1}\mathbf{Q}, \tag{2.63}$$
$$\mathbf{r} = \mathbf{a}(e - \cosh(F)) + \mathbf{b}\sinh(F), \quad e\sinh(F) - F = M, \tag{2.64}$$
$$\mathbf{w} = (-\mathbf{a}\sinh(F) + \mathbf{b}\cosh(F))/(r\sqrt{A}), \tag{2.65}$$
$$\mathbf{v} = \sqrt{\mu}\mathbf{w}. \tag{2.66}$$

2.3 Solving the Basic Kepler's Equation

2.3.1 Iterative Methods

Iteration

Many methods have been suggested for solving this transcendental equation. The simplest of all is the iteration method. One writes Kepler's equation in the form $E = M + e\sin(E)$, sets the initial approximation $E_{old} \approx M$ and repeats the iteration

$$E_{new} = M + e\sin(E_{old}), \tag{2.67}$$

always replacing E_{old} by E_{new} as long as the result changes. This method converges always, but usually slowly and better methods are known.

Newton–Raphson Method

The first of the better methods is Newton's iteration (often called Newton–Raphson iteration). Let E_0 be an approximation. Then

$$E \approx E_0 - (E_0 - e\sin(E_0) - M)/(1 - e\cos(E_0)) \tag{2.68}$$

is an improved approximation (here E_0 and E correspond in meaning to the old and new values above). This correction is to be repeated until the change is negligible. However, the iteration may diverge if $e \approx 1$.

Laguerre's Method

There are methods which always converge. An example is Laguerre's method (Conway, 1986). Write $f = E - e\sin(E) - M$; then $f' = 1 - e\cos(E), f'' = e\sin(E), \ldots$ The iteration

$$E_{n+1} = E_n - f_n/(f'(1-q) + q\sqrt{f_n'^2 - f_n f_n''/q}), \tag{2.69}$$

where $q = 2/3$, gives convergence from the starting value $E_0 = M$. Usually only three iterations are needed, but for $e \approx 1$ and M small, more may be necessary.

This method is actually Laguerre's method for an Nth-order polynomial equation, with $q = 1 - 1/N$. For polynomials it converges always to the nearest (probably complex) root, provided complex algebra is used. For Kepler's equation the best value for q is $2/3$.

The same procedure can be used for the hyperbolic Kepler's equation. Using this method a simple example FORTRAN subroutine for solving Kepler's equation in the elliptic case is as follows:

```
E=ANOM          !  Initial value for E
DO  ITER=1,5
D2=ECC*SIN(E)   ! ECC<1
```

```
      R=1.-ECC*COS(E)
      F=E-D2-ANOM
      DE=-3*F/(R+2*SQRT(R*R-1.5*F*D2))   ! correction
      E=E+DE
      IF(ABS(DE).LT.1.E-7)GO TO 21! error=O(dE^{3})=>1.e-7 sufficient
      END DO
21    CONTINUE
```

More advanced methods for special cases with different solution methods have been published, e.g. by Fukushima (1996, 1997a,b, 1998, 1999, 2008). These, and many others, are too long to be reported here.

Danby and Burkardt Iteration

Danby and Burkardt (1983) and also Burkardt and Danby (1983) suggested a high-order iteration method which is often useful. One writes an equation $f(x+dx) = 0$ in the form $f(x) + dxf'(x) + \frac{1}{2}dx^2 f(x)'' + \frac{1}{6}dx^3 f(x)''' + \cdots = 0$ and solves it for dx as

$$dx_{n+1} = -f(x) \Big/ \sum_{\nu=1}^{n+1} \frac{f^{(\nu)}(x)}{\nu!}(dx_n)^{\nu-1}, \tag{2.70}$$

where dx_n is the nth estimate for the correction dx. This is often efficient because the derivatives must be evaluated only once and the maximum value of n is not large.

2.3.2 Cubic Approximation

There are numerous suggestions for much more advanced methods published in the literature. One of them is the cubic approximation method (Mikkola, 1987).

Let us define a new unknown variable

$$s = \sin(E/3). \tag{2.71}$$

Then we have the relations

$$\sin(E) = 3s - 4s^3 \quad \text{and} \quad E = 3\arcsin(s), \tag{2.72}$$

which lead to the new form of Kepler's equation

$$\arcsin(s) - e(s - \tfrac{4}{3}s^3) = M/3. \tag{2.73}$$

This does not yet give anything new, but if we expand that equation in powers of s to third order, using $\arcsin(s) = s + s^3/6 + O(s^5)$, we have

$$(1-e)s + \left(\frac{1}{6} + \frac{4e}{3}\right)s^3 = \frac{1}{3}M, \tag{2.74}$$

which is an equation of third degree and can be solved with known formulae. After the value of s is found one can obtain the approximation

$$E = M + e(3s - 4s^3), \tag{2.75}$$

which may be improved using the differential correction methods discussed above.

2.3.3 Solving the Cubic Equation

First we consider the use of the well-known Cardano's formula to solve the third-degree equation (2.74). The basic form is not always a good one and thus we present here an alternative that is better for our case.

Let us write

$$p = \frac{(1-e)}{3(\frac{4}{3}e + \frac{1}{6})}, \tag{2.76}$$

$$q = \frac{M}{6(\frac{4}{3}e + \frac{1}{6})}, \tag{2.77}$$

$$z = [\sqrt{p^3 + q^2} + q]^{1/3}, \tag{2.78}$$

$$w = [\sqrt{p^3 + q^2} - q]^{1/3}. \tag{2.79}$$

Then the solution is given by $s = z - w$ which can be better written as

$$s = 2q/(z^2 + p + p^2/z^2) \tag{2.80}$$

and the final approximation for the eccentric anomaly is given by

$$E = M + e(3s - 4s^3). \tag{2.81}$$

The above Cardano's formula for solving cubic equations is not very simple to program and therefore one may consider other alternatives. One way that leads to a simple expression is as follows. Let us write $s = KS$ which, after division by $(1-e)K$, leads to

$$3S + \frac{3(\frac{1}{6} + \frac{4}{3}e)}{(1-e)} K^2 S^3 = \frac{M}{(1-e)K}. \tag{2.82}$$

Now we use the hyperbolic alternative of (2.72)

$$\sinh(3x) = 3\sinh(x) + 4\sinh(x)^3, \tag{2.83}$$

which gives, after setting

$$K = \sqrt{\frac{(1-e)}{(\frac{1}{6} + \frac{4}{3}e)}} \quad \text{and} \quad s = \sinh(x), \tag{2.84}$$

the result

$$\sinh 3x = \frac{M}{(1-e)K}, \tag{2.85}$$

or

$$s = K \sinh\left(\frac{1}{3}\operatorname{arsinh}\left(\frac{M}{(1-e)K}\right)\right). \tag{2.86}$$

A much better approximation can be obtained by replacing the $\frac{1}{6}$ in the above formulae (2.76), (2.77), (2.82) and (2.84) in the expression $(\frac{1}{6} + \frac{4e}{3})$ by

$$\frac{1}{6 - 0.768688675M}, \tag{2.87}$$

in which the constant has been chosen in such a way that the equation is exact at $M = \pi$. This method gives about three correct significant figures in the value of eccentric anomaly E. Thus the final correction with an iterative method is easy, e.g. with the equation (2.70).

2.3.4 Collection of Equations

The set of equations needed to obtain the first approximation for eccentric anomaly is

$$a_6 = 6 - 0.768688675M, \tag{2.88}$$

$$K = \sqrt{\frac{(1-e)}{(1/a_6 + 4e/3)}}, \tag{2.89}$$

$$s = K \sinh\left(\frac{1}{3}\operatorname{arsinh}\left(\frac{M}{(1-e)K}\right)\right), \tag{2.90}$$

$$E = M + e(3s - 4s^3). \tag{2.91}$$

One drawback of this set of equations is that it cannot be used for $e = 1$. However, the equations have a finite limit for $e \to 1$ and in fact very good numerical results can still be obtained by replacing $e = 1$ with a number close to 1, e.g. $e = 1 - 10^{-6}$.

Numerous other methods have been suggested in the literature and the above ones are not always the most efficient. However, the most efficient methods often require long computer programs and they look like 'shooting a fly with a cannon', while the above ones work well and can be programmed with fairly reasonable effort.

2.4 Two-Body Motion in Short Steps

2.4.1 Gauss' f–g Functions

Although the f–g functions were already briefly introduced, a more detailed and self-contained account of these important functions, often called the 'Gaussian f–g functions', is now appropriate.

2.4.2 Elliptic Formulation

Because two-body motion occurs in a plane, one can write

$$\mathbf{r}(t) = \mathbf{r}_0 f(t) + \mathbf{v}_0 g(t), \tag{2.92}$$

where $\mathbf{r}_0 = \mathbf{r}(0)$ and $\mathbf{v}_0 = \mathbf{v}(0)$ are the initial conditions. Clearly $f(0) = 1$ and $g(0) = 0$, and $\dot{f}(0) = 0$, $\dot{g}(0) = 1$, because

$$\mathbf{v}(t) = \mathbf{r}_0 \dot{f}(t) + \mathbf{v}_0 \dot{g}(t). \tag{2.93}$$

Substitution of the f–g representation into $\ddot{\mathbf{r}} = -\mu\mathbf{r}/r^3$ gives

$$\ddot{f} = -\mu f/r^3 \quad \text{and} \quad \ddot{g} = -\mu g/r^3. \tag{2.94}$$

Conservation of angular momentum takes the form

$$f\dot{g} - g\dot{f} = 1. \tag{2.95}$$

Using the elliptical formulae one obtains

$$f = 1 - \frac{a}{r_0}(1 - \cos(\Delta E)), \tag{2.96}$$

$$g = t - a\sqrt{a/\mu}(\Delta E - \sin(\Delta E)), \tag{2.97}$$

$$\dot{f} = \frac{\sqrt{a/\mu}}{r r_0}\sin(\Delta E), \tag{2.98}$$

$$\dot{g} = 1 - \frac{a}{r}(1 - \cos(\Delta E)), \tag{2.99}$$

where $\Delta E = E - E_0$, and E and E_0 are the values of the eccentric anomaly at the moments t and 0. The value of ΔE must be calculated from the difference form of Kepler's equation

$$\Delta E - e(\sin(E_0 + \Delta E) - \sin(E_0)) = \Delta M, \tag{2.100}$$

where $\Delta M = n(t - t_0)$ is the increment of the mean anomaly.

2.4.3 Universal Formulation

To compute f and g in a way independent of the type of orbit (ellipse, parabola or hyperbola), one should employ the *Stumpff functions* $c_\nu(z)$, which may be defined by

$$\cos(x) = c_0(x^2) = 1 - x^2 c_2(x^2) = 1 - x^2/2! + x^4 c_4(x^2) = \cdots,$$
$$\sin(x) = x c_1(x^2) = x - x^3 c_3(x^2) = x - x^3/3! + x^5 c_5(x^2) = \cdots, \tag{2.101}$$

or, generally,

$$c_\nu = \frac{1}{\nu!} - \frac{z}{(\nu+2)!} + \frac{z^2}{(\nu+4)} - + \cdots = \frac{1}{\nu!} - z c_{\nu+2}. \tag{2.102}$$

In addition, it is convenient to introduce the G-functions

$$G_n(\beta, X) = X^n c_n(\beta X^2). \tag{2.103}$$

These G-functions have many useful properties, e.g.

$$\frac{d^2 G_{n+2}}{dX^2} + \beta G_{n+2} = \frac{X^n}{n!}, \tag{2.104}$$

$$\frac{dG_{n+1}}{dX} = G_n, \tag{2.105}$$

$$G_n = \frac{X^n}{n!} - \beta G_{n+2}, \tag{2.106}$$

$$\frac{\partial G_n}{\partial \beta} = \frac{1}{2}(nG_{n+2} - XG_{n+1}). \tag{2.107}$$

The power series for G_n,

$$G_n = \frac{X^n}{n!} - \beta \frac{X^{n+2}}{(n+2)!} + \beta^2 \frac{X^{n+4}}{(n+4)!} - + \cdots, \tag{2.108}$$

is useful for numerical computation if βX^2 is not too large. For large values of the argument, there are special methods (below). By expressing the trigonometric functions with the G-functions (with $z = \beta X^2 = (E - E_0)^2$), one obtains expressions which are also valid for parabolic and hyperbolic motions. From the initial values $\mathbf{r}_0$ and $\mathbf{v}_0$, one computes

$$\beta = 2\mu/r_0 - v^2,$$
$$\eta_0 = \mathbf{r}_0 \cdot \mathbf{v}_0,$$
$$\zeta_0 = \mu - \beta r_0. \tag{2.109}$$

The 'universal' Kepler's equation then takes the form

$$r_0 X + \eta_0 G_2(\beta, X) + \zeta_0 G_3(\beta, X) = t, \tag{2.110}$$

from which $X = \int dt/r$ can be solved. The functions f and g and their derivatives can be written

$$f = 1 - \frac{\mu}{r_0} G_2(\beta, X), \tag{2.111}$$

$$g = t - \mu G_3(\beta, X), \tag{2.112}$$

$$\dot{f} = -\frac{\mu G_1(\beta, X)}{r r_0}, \tag{2.113}$$

$$\dot{g} = 1 - \frac{\mu G_2(\beta, X)}{r}, \tag{2.114}$$

and the position and velocity are given by

$$\mathbf{r} = \mathbf{r}_0 f + \mathbf{v}_0 g, \tag{2.115}$$

$$\mathbf{v} = \mathbf{r}_0 \dot{f} + \mathbf{v}_0 \dot{g}. \tag{2.116}$$

2.4.4 The Universal Kepler's Equation

The universal Kepler's equation (2.110) can be solved using standard methods. Writing

$$\Delta T(X) = r_0 X + \eta_0 G_2(\beta, X) + \zeta_0 G_3(\beta, X) - t \tag{2.117}$$

gives the equation in the form $\Delta T(X) = 0$. The derivatives of this function are

$$\Delta T'(X) = r = r_0 + \eta_0 G_1(\beta, X) + \zeta_0 G_2(\beta, X), \tag{2.118}$$

$$\Delta T''(X) = \eta = \eta_0 G_0(\beta, X) + \zeta_0 G_1(\beta, X), \tag{2.119}$$

$$\Delta T'''(X) = \zeta = \zeta_0 G_0(\beta, X) - \eta_0 \beta G_1(\beta, X), \tag{2.120}$$

which all follow easily from the differentiation and recursion rules for the G-functions. Since the derivatives are easy to evaluate, one can use high-order iteration methods

$$\Delta X_1 = -\Delta T(X)/r, \tag{2.121}$$

$$\Delta X_2 = -\Delta T(X)/(r + \eta(\Delta X_1)^2/2), \tag{2.122}$$

$$\Delta X_3 = -\Delta T(X)/(r + \eta(\Delta X_2)^2/2 + \zeta(\Delta X_2)^3/6), \tag{2.123}$$

$$\vdots$$

obtained by considering the correction ΔX as a solution of

$$\Delta T(X) + r\Delta X + \frac{\eta}{2}(\Delta X)^2 + \frac{\zeta}{6}(\Delta X)^3 = 0. \tag{2.124}$$

An alternative is the Laguerre method

$$\Delta X = -\Delta T(X)/\sqrt{r^2 - \eta \Delta T(X)}. \tag{2.125}$$

When the known approximation for X is good, then the use of the approximations ΔX_1, $\Delta X_2, \ldots$ is efficient, while Laguerre's method has a wider range of convergence.

2.4.5 Kepler's Equation by Bracketing

When solving Kepler's equation, the method of bracketing is almost always successful. The simple principle is as follows:

1. Find an interval in which the solution is found.
2. Use some estimate, e.g. Newton's method, to find an improved estimate for the root of the equation.
3. If the estimate is outside the found interval, then use the mean value of the minimum and value in the interval.
4. Keep always the root bracketed by seeing if the new estimate is smaller or larger than the real root.

In some integration methods, solution for Kepler's equation is needed in equidistant intervals of time (see Chapter 6). In such a situation Kepler's equation should be written for the increment of the solution (like in (2.117)) and a first approximation can be obtained by extrapolation from the previous solutions. Normally quadratic or cubic extrapolation is satisfactory.

Occasionally the solution is more difficult, i.e. when good initial estimates are not available, but in most applications this situation is rare. Usually one wants to have a code that is fast when a good estimate is available, but at the same time it is desirable that the method finds the solution in all cases, including the most difficult ones.

The following method is simple and seems to work always.

Let us consider first the case when solutions are needed at equidistant times. In such a case the Kepler's equation, in Stumpff's form, may be used to move particles over the short equidistant time intervals.

The final solution is then to be obtained using the equation (2.117) which reads

$$\Delta T(X) = r_0 X + \eta_0 G_2(\beta, X) + \zeta_0 G_3(\beta, X) - t = 0. \qquad (2.126)$$

To make the algorithm reliable one may first bracket the solution, i.e. find approximations between which the solution is and use Newton–Raphson iteration from the point that has smallest function value. Alternatively, one may use the equations (2.121), (2.122) and (2.123). If one keeps the solution bracketed, this method usually gives the correct solution even in the most difficult cases. One may choose the first value of X as $X_< = 0$ and the second $X_>$ equal to the extrapolated estimate. If this $X_>$ is too small, i.e. if the values of ΔT have the same sign, one proceeds by computing the value of $\Delta T(X_d)$ as follows:

$$G_0 = 1 - \beta G_2, \tag{2.127}$$

$$\Delta T = r_0 X + \eta_0 G_2 + \zeta_0 G_3 - t, \tag{2.128}$$

$$r_1 = r_0 + \eta_0 G_1 + \zeta_0 G_2, \tag{2.129}$$

$$\eta_1 = \eta_0 G_0 + \zeta_0 G_1, \tag{2.130}$$

$$\zeta_1 = \zeta_0 G_0 - \beta \eta_0 G_1, \tag{2.131}$$

where $G_\nu = G_\nu(\beta, X_d)$ with X_d the increment of X. The first value of X_d can be the extrapolated value of X. One can then repeat the process till bracketing has happened and an interval in which the solution occurs has been found. If it takes too many repetitions then one must simply increase (maybe much) the value of X_d. However, if the new result, obtained e.g. by Newton–Raphson, is outside the bracketed interval, one should bisect the interval, i.e. one takes

$$X_{new} = (X_< + X_>)/2, \tag{2.132}$$

which is then one of the quantities $X_<$ or $X_>$ depending on the sign of $\Delta T(X_{new})$.

Normally it is possible to estimate the solution of the Kepler's equation accurately enough so that troubles are rare. In any case this method is likely to work always even if it is not necessarily the fastest possible.

2.5 Evaluation of c-Functions

Using the trigonometric double-angle formulae, one can derive the following:

$$c_3(z) = \tfrac{1}{4}(c_3(z/4) + c_1(z/4)c_2(z/4)), \tag{2.133}$$

$$c_2(z) = \tfrac{1}{2}c_1(z/4)^2, \tag{2.134}$$

$$c_1(z) = c_1(z/4)c_0(z/4), \tag{2.135}$$

$$c_0(z) = 1 - zc_2(z), \tag{2.136}$$

which may be used to compute the c-functions up to c_3. First one divides the argument by a power of 4, such that a few terms of the series can be used to obtain an accurate value for the functions at that value of the argument. Then the above argument four-folding formulae are used repeatedly until the result at the actual z value is obtained. (Recall that z is the square of the eccentric anomaly increment and thus doubling the angle corresponds to four-folding of z.)

The above formulae are sufficient for computing the motion, but two more of the c-functions are needed to compute partial derivatives and/or differentials of Kepler motion. To evaluate c_4 and c_5, it is possible to use the alternative four-folding formulation:

$$c_5(z) = (c_5(z/4) + c_4(z/4) + c_3(z/4)c_2(z/4))/16, \tag{2.137}$$

$$c_4(z) = c_3(z/4)(1 + c_1(z/4))/8, \tag{2.138}$$

$$c_\nu(z) = 1/\nu! - zc_{\nu+2}(z), \quad \nu = 3,\ 2,\ 1,\ 0. \tag{2.139}$$

2.6 General Algorithms for c-Functions

This section is from Mikkola and Aarseth (1998). Some of the formulae here were earlier given by Waldvogel (1997).

For small argument value the c-functions can be conveniently evaluated by their series expansion. For the purpose of evaluating the c-functions for any argument value we derive a general formula which gives $c(4z)$ once the values are known for argument value z.

The G-functions

$$G_n(\alpha, x) = x^n c_n(\alpha x^2) \tag{2.140}$$

satisfy the differential equation

$$G''_{n+2} + \omega^2 G_{n+2} = \frac{\tau^n}{n!}, \tag{2.141}$$

for $n \geq 0$, with $G_0 = \cos(\omega\tau)$ and $G_1 = \sin(\omega\tau)/\omega$. If we replace τ by $x + t$, we may write

$$G''_{n+2}(x+t) + \omega^2 G_{n+2}(x+t) = \frac{(x+t)^n}{n!} = \sum_{\nu=0}^{n} \frac{t^{n-\nu}}{(n-\nu)!} \frac{x^\nu}{\nu!}. \tag{2.142}$$

Thus, with the help of the previous differential equation, we see that the solution can also be written as

$$G_{n+2}(x+t) = G_{n+2}(t)G_0(x) + G_{n+1}(t)G_1(x) + \sum_{\nu=0}^{n} \frac{t^{n-\nu}}{(n-\nu)!} G_{\nu+2}(x). \tag{2.143}$$

If we substitute $t = -x$ and use the definition (2.140), we obtain the relation

$$0 = c_{n+2}c_0 - c_{n+1}c_1 + \sum_{\nu=0}^{n} \frac{(-1)^{\nu}}{(n-\nu)!} c_{\nu+2}, \tag{2.144}$$

and by setting $t = x$ we have

$$2^{n+2} c_{n+2}(4z) = c_{n+2}c_0 + c_{n+1}c_1 + \sum_{\nu=0}^{n} \frac{1}{(n-\nu)!} c_{\nu+2}, \tag{2.145}$$

where the c-functions without explicit argument are understood as $c(z)$. The latter of these equations already gives an expression with which the four-fold argument function may be evaluated. However, if we add or subtract the previous equation from this, a shorter expression follows. The two possibilities are

$$c_{n+2}(4z) = \frac{1}{2^{n+1}} \left(c_{n+2}c_0 + \sum_{\nu=0,2,n} \frac{c_{\nu+2}}{(n-\nu)!} \right) \tag{2.146}$$

and

$$c_{n+2}(4z) = \frac{1}{2^{n+1}} \left(c_{n+1}c_1 + \sum_{\nu=1,2,n} \frac{c_{\nu+2}}{(n-\nu)!} \right). \tag{2.147}$$

Here the summation signs written as $\sum_{\nu=0,2,n}$ and $\sum_{\nu=1,2,n}$ mean that the summation index is incremented by 2 until it reaches at most the value of n.

The algorithm for evaluation of c-functions thus takes the form:

1. First divide z by a power of 4 (say 4^k) to obtain $z' = z/4^k$ so small that the series expansion can be used to obtain the two highest-order functions required at this reduced value z'.
2. Apply the recursion $c_n = 1/n! - z'c_{n+2}$ to obtain the lower-order functions up to c_0.
3. Finally use the above four-fold argument formula for the two highest-order functions. Replace z' by $4z'$ and compute the lower-order ones by the above recursion.
4. Thus one has computed the c-functions at the argument value of $4z'$.
5. Repeat this operation k times to reach the original value of z.

A Simple and Accurate c-Function Routine in FORTRAN

```
SUBROUTINE cfun(z,c)!Stumpff(Z,C)
IMPLICIT REAL*8 (A-H,O-Z)
parameter(o2=1.d0/2,o6=1.d0/6,o8=1.d0/8,o16=1.d0/16)
Real*8 C(0:5)
save
```

```
      h=z
      DO  K=0,7
      IF(ABS(h).lt.0.9d0)goto 2
      h=h/4 ! divide by 4 untill h<.9
      END DO
2     C(4)= (201859257600.d0+h*(-3741257520.d0  ! Pade 4 c4
     & +(40025040.d0-147173.d0*h)*h))/
     & (240.d0*(20185925760.d0 + h*(298738440.d0
     & + h*(1945020.d0 + 5801.d0*h))))
      C(5)=(3750361655040.d0 + h*(-40967886960.d0 ! Pade 4 c5
     & + (358614256.d0 - 1029037.d0*h)*h))/
     & (55440.d0*(8117665920.d0 + h*(104602680.d0
     &    + h*(582348.d0 + 1451.d0*h))))
      DO  I=1,K  ! 4-fold argument K times
      C3=o6-h*C(5)
      C2=o2-h*C(4)
      C(5)=(C(5)+C(4)+C2*C3)*o16
      C(4)=C3*(2.D0-h*C3)*o8
      h=4.d0*h
      END DO
      C(3)=o6-Z*C(5)
      C(2)=o2-Z*C(4)
      C(1)=1-Z*C(3)
      C(0)=1-Z*C(2)
      RETURN
      END
```

Chapter 3
Analytical Tools

3.1 The Lagrangian

The basic variational principle in classical mechanics is the Hamiltonian principle, which is discussed in, for example, Hand and Finch (2008) and Goldstein, Poole and Safko (2002) and especially in Hagihara (1970). The integral

$$I = \int_{t_1}^{t_2} L\,dt, \tag{3.1}$$

where $L = T - V$ is the difference of the kinetic and potential energies, has an extremum value along the trajectory. (Note: In what follows we often use, instead of the potential energy V, the force function $U = -V$.)

If $L = L(\mathbf{q}, \dot{\mathbf{q}}, t)$, where the $\mathbf{q}$ are generalized coordinates and the $\dot{\mathbf{q}}$ their derivatives, then the extremum condition gives the *Lagrange equations*

$$\frac{d}{dt}\frac{\partial L}{\partial \dot{\mathbf{q}}} - \frac{\partial L}{\partial \mathbf{q}} = 0. \tag{3.2}$$

(In the calculus of variations these are known as the Euler equations.)

Example: Two-body equations. If m_a and m_b are the masses, $\mathbf{v}_a$ and $\mathbf{v}_b$ the velocities and $\mathbf{r}_a$ and $\mathbf{r}_b$ the position vectors, one may write

kinetic energy $= \frac{1}{2}m_a\mathbf{v}_a^2 + \frac{1}{2}m_b\mathbf{v}_b^2$,

force $= \mathbf{F}_{ab} = Gm_am_b(\mathbf{r}_a - \mathbf{r}_b)/r_{ab}^3 = Gm_am_b\dfrac{\partial}{\partial \mathbf{r}_a}\dfrac{1}{r_{ab}}$

and

$$\mathbf{F}_{ab} = Gm_a m_b \frac{\partial}{\partial \mathbf{r}_b} \frac{1}{r_{ab}}.$$

Thus the force function is

$$U = G\frac{m_a m_b}{r_{ab}}$$

and the Lagrangian reads

$$L = \frac{1}{2}m_a \mathbf{v}_a^2 + \frac{1}{2}m_b \mathbf{v}_b^2 + G\frac{m_a m_b}{r_{ab}}. \tag{3.3}$$

Then

$$\frac{\partial L}{\partial \mathbf{v}_a} = m_a \dot{\mathbf{v}}_a, \qquad \frac{\partial L}{\partial \mathbf{v}_b} = m_b \mathbf{v}_b \tag{3.4}$$

and

$$\frac{\partial L}{\partial \mathbf{r}_a} = Gm_a m_b(\mathbf{r}_b - \mathbf{r}_a)/r_{ab}^3, \qquad \frac{\partial L}{\partial \mathbf{r}_b} = Gm_a m_b(\mathbf{r}_b - \mathbf{r}_a)/r_{ab}^3 \tag{3.5}$$

so that

$$\frac{d}{dt}\frac{\partial L}{\partial \mathbf{v}_a} = m_a \ddot{\mathbf{r}}_a = \frac{\partial L}{\partial \mathbf{r}_a} = Gm_a m_b \frac{\mathbf{r}_b - \mathbf{r}_a}{r_{ab}^3} \tag{3.6}$$

and

$$\ddot{\mathbf{r}}_a = Gm_b(\mathbf{r}_b - \mathbf{r}_a)/r_{ab}^3, \tag{3.7}$$

$$\ddot{\mathbf{r}}_b = Gm_a(\mathbf{r}_a - \mathbf{r}_b)/r_{ab}^3. \tag{3.8}$$

3.2 The Hamiltonian

Define the *Hamiltonian*

$$H = \sum \dot{q}_k p_k - L(\mathbf{q}, \dot{\mathbf{q}}, t), \tag{3.9}$$

where p_k are the *momenta*,

$$p_k = \frac{\partial L}{\partial \dot{q}_k}. \tag{3.10}$$

Now

$$\frac{\partial H}{\partial p_k} = \dot{q}_k \quad \text{and} \quad \frac{\partial H}{\partial q_k} = -\frac{\partial L}{\partial q_k} = -\frac{d}{dt}\frac{\partial L}{\partial \dot{q}_k} = -\frac{dp_k}{dt}$$

or

$$\dot{q}_k = \frac{\partial H}{\partial p_k}, \tag{3.11}$$

$$\dot{p}_k = -\frac{\partial H}{\partial q_k}. \tag{3.12}$$

In addition

$$\frac{\partial H}{\partial t} = -\frac{\partial L}{\partial t}. \tag{3.13}$$

These are the Hamiltonian equations of motion in canonical form. The quantities p_k and q_k are canonically conjugate variables (momenta p and coordinates q). If the kinetic energy T is a homogeneous quadratic form

$$T = \sum A_{ij}\dot{q}_i\dot{q}_j, \tag{3.14}$$

then $H = T - U = E =$ energy, for

$$\begin{aligned} H &= \sum_k \dot{q}_k \frac{\partial T}{\partial \dot{q}_k} - L = 2T - L = 2T - (T - U) \\ &= T - U = T + V = E. \end{aligned} \tag{3.15}$$

Example: Two-body problem. The momenta are

$$\mathbf{p}_a = \frac{\partial T}{\partial \mathbf{v}_a} = m_a\mathbf{v}_a, \quad \mathbf{p}_b = \frac{\partial T}{\partial \mathbf{v}_b} = m_b\mathbf{v}_b \tag{3.16}$$

and the kinetic and potential energies take the form

$$T = \frac{p_a^2}{2m_a} + \frac{p_b^2}{2m_b}, \quad U = G\frac{m_a m_b}{r_{ab}}. \tag{3.17}$$

Thus, using $H = T - U$, one obtains

$$\dot{\mathbf{r}}_a = \frac{\partial H}{\partial \mathbf{p}_a} = \frac{\mathbf{p}_a}{m_a}(= \mathbf{v}_a) \quad \text{and} \quad \dot{\mathbf{p}}_a = -\frac{\partial H}{\partial \mathbf{r}_a} = G m_a m_b \frac{\mathbf{r}_b - \mathbf{r}_a}{r_{ab}^3} \tag{3.18}$$

or

$$\ddot{\mathbf{r}}_a = \frac{\dot{\mathbf{p}}_a}{m_a} = G m_b \frac{\mathbf{r}_b - \mathbf{r}_a}{r_{ab}^3} \tag{3.19}$$

(and similarly for $\mathbf{r}_b$), as expected.

3.3 Canonical Transformations

A transformation

$$\mathbf{Q} = \mathbf{Q}(\mathbf{q}, \mathbf{p}), \quad \mathbf{P} = \mathbf{P}(\mathbf{q}, \mathbf{p}) \tag{3.20}$$

is called a *canonical transformation* if there exists a new Hamiltonian function $\widetilde{H}(\mathbf{P}, \mathbf{Q})$, depending on the new coordinates $\mathbf{Q}$ and momenta $\mathbf{P}$, such that the equations of motion for $\mathbf{P}$ and $\mathbf{Q}$ are canonical in form, i.e. $\dot{\mathbf{P}} = -\partial \widetilde{H}/\partial \mathbf{Q}$ and $\dot{\mathbf{Q}} = \partial \widetilde{H}/\partial \mathbf{P}$.

The *point transformation* is one in which the coordinate transformation has the form $\mathbf{Q} = \mathbf{Q}(\mathbf{q})$. The transformation formula for the momenta can be obtained by means of a *generating function* S, where

$$S = \mathbf{P} \cdot \mathbf{Q}(\mathbf{q}). \tag{3.21}$$

Then

$$\mathbf{p} = \frac{\partial S}{\partial \mathbf{q}} \tag{3.22}$$

and

$$\widetilde{H} = H(\mathbf{p}(\mathbf{Q}, \mathbf{P}), \mathbf{q}(\mathbf{P}, \mathbf{Q})). \tag{3.23}$$

If the transformation is given as $q = q(Q)$, one may use

$$S = \mathbf{p} \cdot \mathbf{q}(\mathbf{Q}) \tag{3.24}$$

and

$$\mathbf{P} = \frac{\partial S}{\partial \mathbf{Q}}. \tag{3.25}$$

Example: Polar coordinates. Let $H = \frac{1}{2}(p_x^2 + p_y^2) - 1/\sqrt{x^2 + y^2}$.
Transformation: $x = r\cos(\theta)$, $y = r\sin(\theta)$.
Generating function: $S = p_S r\cos(\theta) + p_y r\sin(\theta)$.
Momentum transformation: $P_r = \partial S/\partial r = p_x \cos(\theta) + p_y \sin(\theta)$ and $P_\theta = \partial S/\partial \theta = r(-p_x \sin(\theta) + p_y \cos(\theta))$.
Thus the new Hamiltonian is $\widetilde{H} = \frac{1}{2}(P_r^2 + P_\theta^2/r^2) - 1/r$.

More Canonical Transformations

Without going deeper into the general theory, we list here some generating functions for canonical transformations:

$$1. \quad S = S_1(q, Q, t), \tag{3.26}$$

$$2. \quad S = S_2(q, P, t), \tag{3.27}$$

$$3. \quad S = S_3(p, Q, t), \tag{3.28}$$

$$4. \quad S = S_4(p, P, t). \tag{3.29}$$

Here lower-case letters signify 'old' variables and upper-case letters represent the 'new' variables. Transformation formulae for the above generating functions read:

1. $p = \partial S/\partial q, \quad P = -\partial S/\partial Q, \quad H_{new} = H + \partial S/\partial t,$
2. $P = \partial S/\partial q, \quad Q = \partial S/\partial P, \quad H_{new} = H + \partial S/\partial t,$
3. $q = -\partial S/\partial p, \quad P = -\partial S/\partial Q, \quad H_{new} = H + \partial S/\partial t,$
4. $q = -\partial S/\partial p, \quad Q = \partial S/\partial P, \quad H_{new} = H + \partial S/\partial t.$

These may be derived from the general result. The Lagrangians corresponding to the new and old Hamiltonians are

$$L_{new} = \sum P\dot{Q} - H_{new}(Q, P, t), \tag{3.30}$$

$$L_{old} = \sum p\dot{q} - H(p, q, t). \tag{3.31}$$

From variational calculus, the equations

$$\delta \int L_{new} dt = 0, \quad \delta \int L_{old} dt = 0 \tag{3.32}$$

imply that the L differ by the time derivative $\partial S/\partial t$ of an arbitrary function S.

An important class of canonical transformations is defined by Hamiltonian equations. Let $\mathcal{X} = \mathcal{X}(\mathbf{p}, \mathbf{q})$ be an arbitrary function. If we take $\mathcal{X}$ to be a Hamiltonian and solve the corresponding equations of motion followed by an advancement of the solution over a chosen interval of time, then the transformation $(\mathbf{p}_0, \mathbf{q}_0) \to (\mathbf{p}(t), \mathbf{q}(t))$ is a canonical transformation. In particular, by taking $\mathcal{X} = H$, we realize that the motion of a Hamiltonian system is a canonical transformation.

3.4 Time-Dependent Hamiltonian

Let

$$H = H(p_1, p_2, \ldots, p_n, q_1, q_2, \ldots, q_n, t). \tag{3.33}$$

Consider time as a coordinate $= q_0$ and introduce the corresponding conjugate momentum p_0 as well as the new Hamiltonian (Stiefel and Scheifele, 1971)

$$\widetilde{H} = H + p_0, \tag{3.34}$$

which gives the correct equations of motion $\dot{p}_k = -\partial\widetilde{H}/\partial q_k$ and $\dot{q}_k = \partial\widetilde{H}/\partial p_k$ for $k = 1, 2, \ldots, n$. For $k = 0$, we obtain

$$\dot{q}_0 = \frac{\partial \widetilde{H}}{\partial p_0} = 1 \quad \text{or} \quad q_0 = t \tag{3.35}$$

and

$$\dot{p}_0 = -\frac{\partial \widetilde{H}}{\partial q_0} = -\frac{\partial H}{\partial q_0} = -\frac{\partial H}{\partial t}. \tag{3.36}$$

Our new Hamiltonian $\widetilde{H}$ is constant along the orbit, because

$$\frac{d\widetilde{H}}{dt} = \sum_{k=0}^{n} \left(\frac{\partial \widetilde{H}}{\partial q_k} \dot{q}_k + \frac{\partial \widetilde{H}}{\partial p_k} \dot{p}_k \right) = \sum_{k=0}^{n} \left(\frac{\partial \widetilde{H}}{\partial q_k} \frac{\partial \widetilde{H}}{\partial p_k} - \frac{\partial \widetilde{H}}{\partial p_k} \frac{\partial \widetilde{H}}{\partial q_k} \right) = 0. \tag{3.37}$$

Consequently $\widetilde{H} = H + p_0$ is always a constant of motion and $dH/dt = -dp_0/dt = \partial H/\partial t$. Also, if H does not depend explicitly on time, it is a constant of motion. If initially

$$p_0(0) = -H(0), \tag{3.38}$$

then *along the trajectory*

$$\widetilde{H}(t) = 0. \tag{3.39}$$

3.5 Time-Dependent Point Transformation

When the Hamiltonian is

$$\widetilde{H} = H(p_1, \ldots, p_n, q_1, \ldots, q_n, q_0) + p_0 \tag{3.40}$$

and the generating function depends on time

$$S = S(\ldots, q_0), \tag{3.41}$$

then the transformation formula for p_0 is

$$p_0 = \frac{\partial S}{\partial q_0}. \tag{3.42}$$

Consequently the transformed Hamiltonian is $\widetilde{\widetilde{H}} = H(\ldots, q_0) + \partial S/\partial q_0$. Thus, if S depends on time, the term $\partial S/\partial t$ must be added to the new Hamiltonian, i.e.

$$\widetilde{\widetilde{H}} = H(\ldots, t) + \frac{\partial S}{\partial t}, \quad (\mathbf{p} = \mathbf{p}(\mathbf{P}, \mathbf{Q}, t),\ \mathbf{q} = \mathbf{q}(\mathbf{P}, \mathbf{Q}, t)). \tag{3.43}$$

Example: Rotating coordinates.
Hamiltonian: $H = \frac{1}{2}(p_x^2 + p_y^2) - U(x, y, t)$.
Transformation: $Q_1 = x\cos(t) + y\sin(t)$ and $Q_2 = -x\sin(t) + y\cos(t)$.
Thus: $S = P_1 Q_1 + P_2 Q_2$, $p_x = \partial S/\partial x = P_1\cos(t) - P_2\sin(t)$ and $p_y = \partial S/\partial y = P_1\sin(t) + P_2\cos(t) \Longrightarrow p_x^2 + p_y^2 = P_1^2 + P_2^2$,
$\partial S/\partial t = P_1(-x\sin(t) + y\cos(t)) + P_2(-x\cos(t) - y\sin(t)) = P_1 Q_2 - P_2 Q_1$.
Thus the Hamiltonian in rotating coordinates reads
$\widetilde{H} = \frac{1}{2}(P_1^2 + P_2^2) - U(Q_1, Q_2, t) + P_1 Q_2 - P_2 Q_1$.
This is a useful transformation if U happens to be time-independent in the rotating system, i.e. if $U(Q_1\cos(t) - Q_2\sin(t), Q_1\sin(t) +$

$Q_2 \cos(t), t) = \widetilde{U}(Q_1, Q_2)$, in which case the new Hamiltonian is a constant of motion. This is the case in the so-called restricted three-body problem.

3.6 Poincaré's Time Transformation

If, as the independent variable, a new variable s is introduced such that $t = t(s)$, then the Hamiltonian equations are replaced by

$$\mathbf{p}' = -t' \frac{\partial H}{\partial \mathbf{q}} \quad \text{and} \quad \mathbf{q}' = t' \frac{\partial H}{\partial \mathbf{p}}, \tag{3.44}$$

where the prime indicates differentiation with respect to s. In general we may, instead of $t(s)$, choose a function $t' = g(\mathbf{p}, \mathbf{q})$ (or $dt = g(\mathbf{p}, \mathbf{q}) ds$). In this case

$$\mathbf{p}' = -g(\mathbf{p}, \mathbf{q}) \frac{\partial H}{\partial \mathbf{q}}, \quad \mathbf{q}' = g(\mathbf{p}, \mathbf{q}) \frac{\partial H}{\partial \mathbf{p}}. \tag{3.45}$$

These equations are in the canonical form. It is, however, possible to convert the equations into the Hamiltonian form in this case too. This can be accomplished by using, instead of H, the Poincaré function (Poincaré, 1895)

$$\Gamma = g(\mathbf{p}, \mathbf{q})(H + p_0) \tag{3.46}$$

and choosing $p_0(0) = -H(0) = E_0$. Along the orbit $H + p_0 = 0$ and

$$\frac{\partial \Gamma}{\partial \mathbf{p}} = g \frac{\partial H}{\partial \mathbf{p}} + \frac{\partial g}{\partial \mathbf{p}} (H + p_0) = g \frac{\partial H}{\partial \mathbf{p}} = \mathbf{q}', \tag{3.47}$$

$$\frac{\partial \Gamma}{\partial \mathbf{q}} = g \frac{\partial H}{\partial \mathbf{q}} + \frac{\partial g}{\partial \mathbf{q}} (H + p_0) = g \frac{\partial H}{\partial \mathbf{q}} = -\mathbf{p}', \tag{3.48}$$

$$\frac{\partial \Gamma}{\partial p_0} = g = t', \tag{3.49}$$

$$\frac{\partial \Gamma}{\partial q_0} = g \frac{\partial H}{\partial t} = -p_0'. \tag{3.50}$$

Thus the new equations are canonical in form. Note, however, that $H + p_0$ is not identically zero but it only has this value on the correct orbit. A consequence of this is that one must not drop the terms factored by $H + p_0$ because then the Hamiltonian form would be lost.

Example: One-dimensional two-body problem. $H = \frac{1}{2}p^2 - 1/r$.
Transformation: $r = Q^2$, $S = pQ^2$, $P = \partial S / \partial r = 2pQ$, $p = P/(2Q)$ and $H = P^2/(8Q^2) - 1/Q^2$.
Poincaré's function is $\Gamma = g(H + p_0) = g(P^2/8Q^2 - 1/Q^2 - E_0)$ (because p_0 is constant, we must take $p_0 = -E_0$).

The choice $g = r = Q^2$ simplifies the new Hamiltonian
$\Gamma = \frac{1}{8}P^2 - 1 - E_0 Q^2$,
which is the Hamiltonian of the harmonic oscillator and the solution is $Q = Q(0)\cos(\omega s) + Q'(0)\sin(\omega s)/\omega$, where $\omega = \sqrt{-E_0/2}$.
In addition one must solve the differential equation for time $t' = Q^2$, which gives $t = \int (Q(0)\cos(\omega s) + Q'(0)\sin(\omega s)/\omega)^2 ds$. This is easy to integrate and actually gives what is known as the Kepler's equation.

3.7 Hamilton–Jacobi Equation

If a transformation can be found which makes the new Hamiltonian identically equal to zero, the new coordinates and momenta are constants of motion. This means then that the equations of motion have been solved via the transformation. For a generating function of the form $S = S(\mathbf{p}, \mathbf{q})$ the transformations are $\mathbf{p} = \partial S/\partial \mathbf{q}$ and $H(\mathbf{p}, \mathbf{q}) + \partial S/\partial t = 0$, and we get the *Hamilton–Jacobi (HJ) equation*

$$H\left(\frac{\partial S}{\partial \mathbf{q}}, \mathbf{q}\right) + \frac{\partial S}{\partial t} = 0. \tag{3.51}$$

The solution of this partial differential equation thus also solves Hamiltonian equations of motion. Unfortunately, solving this HJ equation is usually just as difficult as the solution of the equations of motion. The equation can usually be applied only if one finds coordinates $\mathbf{q}$ in which the system becomes separable such that it is possible to write

$$S = \sum_k S(q_k). \tag{3.52}$$

Problems in which this is possible include the two-body motion.

Chapter 4
Variation of Parameters

4.1 Lagrange's Formulation

The solution of the unperturbed two-body equation of motion

$$\ddot{\mathbf{r}} = -\mu\mathbf{r}/r^3, \quad (\mu = G(m_1 + m_2)) \tag{4.1}$$

is called *pure Kepler motion*. A Keplerian orbit in which the position and velocity instantaneously coincide with the actual values is called the *osculating orbit*. If the osculating orbit is known, one may evaluate the position and velocity using Keplerian-motion formulae. If the motion is perturbed, the osculating orbit changes slowly (if the perturbation is small). Perturbation theory attempts to determine the osculating orbit as a function of time. In most of the considerations below it does not matter whether the perturbing force has a potential or not. However, if canonical equations are employed, it is assumed that the Hamiltonian has the form

$$H = K - R = \tfrac{1}{2}\mathbf{p}^2 - \mu/r - R, \tag{4.2}$$

where R is the *perturbing function*. All equations are, however, still valid unless the perturbing function itself appears in the expressions.

Let the perturbed equation of motion be

$$\ddot{\mathbf{r}} = -\mu\mathbf{r}/r^3 + \mathbf{F}. \tag{4.3}$$

If the solution is presented in terms of an osculating motion, the (two-body) constants of motion c_i must be functions of time:

$$\mathbf{r}(t) = \mathbf{r}(c_1(t), c_2(t), \ldots, c_6(t), t). \tag{4.4}$$

The osculation condition is

$$\frac{d\mathbf{r}}{dt} = \frac{\partial \mathbf{r}}{\partial t}. \tag{4.5}$$

On the other hand

$$\frac{d\mathbf{r}}{dt} = \frac{\partial \mathbf{r}}{\partial t} + \sum_{\nu=1}^{6} \frac{\partial \mathbf{r}}{\partial c_\nu} \dot{c}_\nu, \tag{4.6}$$

so that

$$\sum_{\nu=1}^{6} \frac{\partial \mathbf{r}}{\partial c_\nu} \dot{c}_\nu = \mathbf{0}. \tag{4.7}$$

Let the velocity, computed by the two-body formulae, be

$$\mathbf{v} = \frac{\partial \mathbf{r}}{\partial t} = \mathbf{v}(c_1(t),\ c_2(t), \ldots,\ c_6(t), t). \tag{4.8}$$

In osculating motion, this is equal to the true velocity. From the equation of pure Keplerian motion,

$$\frac{\partial \mathbf{v}}{\partial t} = -\mu \mathbf{r}/r^3, \tag{4.9}$$

so that, with $\ddot{\mathbf{r}} = \dot{\mathbf{v}}$,

$$\frac{d\mathbf{v}}{dt} = \frac{\partial \mathbf{v}}{\partial t} + \sum_{\nu=1}^{6} \frac{\partial \mathbf{v}}{\partial c_\nu} \dot{c}_\nu = -\mu \mathbf{r}/r^3 + \mathbf{F}. \tag{4.10}$$

By combining these results with the osculation condition, one obtains

$$\sum_{\nu=1}^{6} \frac{\partial \mathbf{v}}{\partial c_\nu} \dot{c}_\nu = \mathbf{F}, \tag{4.11}$$

$$\sum_{\nu=1}^{6} \frac{\partial \mathbf{r}}{\partial c_\nu} \dot{c}_\nu = 0, \tag{4.12}$$

where there exist the necessary number of equations for the determination of the time derivatives of the 'constants' c_ν. Scalar multiplication of the first equation with $\partial \mathbf{r}/\partial c_\mu$ and the second with $\partial \mathbf{v}/\partial c_\mu$, followed by subtraction of the results, gives

$$\sum_{\nu} \left\{ \frac{\partial \mathbf{r}}{\partial c_\mu} \cdot \frac{\partial \mathbf{v}}{\partial c_\nu} - \frac{\partial \mathbf{v}}{\partial c_\mu} \cdot \frac{\partial \mathbf{r}}{\partial c_\nu} \right\} \dot{c}_\nu = \mathbf{F} \cdot \frac{\partial \mathbf{r}}{\partial c_\mu}. \tag{4.13}$$

The bracketed expression here is usually called the *Lagrange bracket* and is often abbreviated $\{c_\mu, c_\nu\}$. These brackets do not depend explicitly on time and are thus constants in pure Keplerian motion. An important consequence is that *the moment of time may be chosen conveniently* to evaluate the values of these quantities.

One may take the (parametric) initial values $\mathbf{r}_0$ and $\mathbf{v}_0$ of the osculating Kepler orbit, at an arbitrary moment t_0, as the constants of motion

$$(c_1, c_2, c_3, c_4, c_5, c_6) = (x_0, y_0, z_0, \dot{x}_0, \dot{y}_0, \dot{z}_0),$$

and then conveniently evaluate the Lagrange brackets by choosing the moment of time to be $t = t_0$. Then

$$\{x_0, \dot{x}_0\} = \overbrace{\underbrace{\frac{\partial \mathbf{r}_0}{\partial x_0}}_{(1,0,0)} \cdot \underbrace{\frac{\partial \mathbf{v}_0}{\partial \dot{x}_0}}_{(1,0,0)}}^{=1} - \overbrace{\underbrace{\frac{\partial \mathbf{v}_0}{\partial x_0}}_{=\mathbf{0}} \cdot \underbrace{\frac{\partial \mathbf{r}_0}{\partial \dot{x}_0}}_{=\mathbf{0}}}^{=0} = 1 = \{y_0, \dot{y}_0\} = \{z_0, \dot{z}_0\}. \tag{4.14}$$

Because $\{c_\mu, c_\nu\} = -\{c_\nu, c_\mu\}$, we have $\{\dot{x}_0, x_0\} = -1 = \{\dot{y}_0, y_0\} = \{\dot{z}_0, z_0\}$, and all the other brackets are zeros. Thus the matrix of Lagrange brackets is diagonal, which leads to the simple solutions $d\dot{x}_0/dt = \mathbf{F} \cdot \partial \mathbf{r}(t)/\partial x_0 = \partial R/\partial x_0$ and $dx_0/dt = -\mathbf{F} \cdot \partial \mathbf{r}(t)/\partial \dot{x}_0 = -\partial R/\partial \dot{x}_0$. It also follows that the parametric initial conditions are canonical elements, so that

$$\frac{d\mathbf{v}_0}{dt} = \frac{\partial R}{\partial \mathbf{r}_0}, \quad \frac{d\mathbf{r}_0}{dt} = -\frac{\partial R}{\partial \mathbf{v}_0}. \tag{4.15}$$

4.2 Poisson's Equations

It is possible to obtain the solution for the Lagrange perturbation equations 'directly'. From the equations

$$\mathbf{r} = \mathbf{r}(c_1, \ldots, c_6, t), \tag{4.16}$$

$$\mathbf{v} = \mathbf{v}(c_1, \ldots, c_6, t), \tag{4.17}$$

the constants c_ν are solved in terms of $\mathbf{r}$ and $\mathbf{v}$:

$$c_\nu = c_\nu(\mathbf{r}, \mathbf{v}, t). \tag{4.18}$$

Differentiation with respect to the time gives

$$\begin{aligned} \frac{dc_\nu}{dt} &= \frac{\partial c_\nu}{\partial \mathbf{r}} \cdot \frac{d\mathbf{r}}{dt} + \frac{\partial c_\nu}{\partial \mathbf{v}} \cdot \frac{d\mathbf{v}}{dt} + \frac{\partial c_\nu}{\partial t} \\ &= \frac{\partial c_\nu}{\partial \mathbf{r}} \cdot \frac{\partial \mathbf{r}}{\partial t} + \frac{\partial c_\nu}{\partial \mathbf{v}} \cdot \frac{\partial \mathbf{v}}{\partial t} + \frac{\partial c_\nu}{\partial t} + \frac{\partial c_\nu}{\partial \mathbf{v}} \cdot \mathbf{F}, \end{aligned} \tag{4.19}$$

where the osculation condition

$$\frac{d\mathbf{r}}{dt} = \frac{\partial \mathbf{r}}{\partial t}, \quad \frac{d\mathbf{v}}{dt} = \frac{\partial \mathbf{v}}{\partial t} + \mathbf{F}$$

has been used. In addition, as c_ν is constant in the pure Keplerian motion,

$$\frac{\partial c_\nu}{\partial \mathbf{r}} \cdot \frac{\partial \mathbf{r}}{\partial t} + \frac{\partial c_\nu}{\partial \mathbf{v}} \cdot \frac{\partial \mathbf{v}}{\partial t} + \frac{\partial c_\nu}{\partial t} = 0. \tag{4.20}$$

By subtracting this from the previous equation we get the simple result

$$\frac{dc_\nu}{dt} = \frac{\partial c_\nu}{\partial \mathbf{v}} \cdot \mathbf{F}, \tag{4.21}$$

which is Poisson's equation for perturbation.

4.3 Herrick's Perturbative Variation

Consider an arbitrary function of time, position and velocity $f = f(t, \mathbf{r}, \mathbf{v})$. This can be converted into the form

$$f = f(t, c_1, \ldots, c_6), \tag{4.22}$$

which has the time derivative

$$\begin{aligned} \frac{df}{dt} &= \frac{\partial f}{\partial t} + \sum \frac{\partial f}{\partial c_\nu} \dot{c}_\nu = \frac{\partial f}{\partial t} + \overbrace{\sum \frac{\partial f}{\partial c_\nu} \frac{\partial c_\nu}{\partial \mathbf{v}}}^{(\partial f/\partial \mathbf{v})} \cdot \mathbf{F} \\ &= \frac{\partial f}{\partial t} + \frac{\partial f}{\partial \mathbf{v}} \cdot \mathbf{F}. \end{aligned} \tag{4.23}$$

Here the first term $\partial f/\partial t$ is the derivative which f would have in pure Keplerian motion and the latter part is due to the perturbation and was called by Herrick (1972) the *perturbative variation*. It is often abbreviated $f^{\grave{}}$. Generally one may write

$$\frac{df}{dt} = \frac{\partial f}{\partial t} + f^{\grave{}}, \tag{4.24}$$

where $f^{\grave{}} = (\partial f/\partial \mathbf{v}) \cdot \mathbf{F}$. It is important to realize that

$$\mathbf{v}^{\grave{}} = \mathbf{F} \quad \text{and} \quad \mathbf{r}^{\grave{}} = \mathbf{0}. \tag{4.25}$$

The motivation for the introduction of $f^{\grave{}}$ is mainly that, for a constant of Keplerian motion,

$$\frac{dc_\nu}{dt} = c_\nu^{\grave{}}, \tag{4.26}$$

and that the perturbative variation is a differential operator. If R is an arbitrary function $R = R(t, f_1, \ldots, f_n)$, then

$$R^{\grave{}} = \frac{\partial R}{\partial \mathbf{v}} \cdot \mathbf{F} = \sum_k \frac{\partial R}{\partial f_k} \frac{\partial f_k}{\partial \mathbf{v}} \cdot \mathbf{F} = \sum_k \frac{\partial R}{\partial f_k} f_k^{\grave{}}, \tag{4.27}$$

which means that *all the rules of normal differentiation apply to the perturbative differentiation.* Therefore the time derivatives of Keplerian motion elements can be obtained by differentiating (in the perturbative sense) the formulae (equation by equation) for the computation of those elements. One must only take into account that $\mathbf{r}^{`} = \mathbf{0}$ and $\mathbf{v}^{`} = \mathbf{F}$.

Examples:

1. Inverse semi-major axis $\alpha = 2/r - \mathbf{v}^2/\mu$, thus
$$\frac{d\alpha}{dt} = \alpha^{`} = \frac{\partial\alpha}{\partial\mathbf{v}} \cdot \mathbf{F} = -2\mathbf{v} \cdot \mathbf{F}/\mu.$$
2. Area velocity $\mathbf{A} = \mathbf{r} \times \mathbf{v}$, thus
$$\frac{d\mathbf{A}}{dt} = \mathbf{A}^{`} = \mathbf{r} \times \mathbf{v}^{`} = \mathbf{r} \times \mathbf{F}.$$
3. Eccentricity $e^2 = (1 - r\alpha)^2 + \alpha(\mathbf{r} \cdot \mathbf{v})^2/\mu$, thus
$$2ee^{`} = ((-2(1 - r\alpha)r + (\mathbf{r} \cdot \mathbf{v})^2/\mu)\alpha^{`} + 2\alpha(\mathbf{r} \cdot \mathbf{v})\mathbf{r} \cdot \mathbf{F}/\mu).$$

4.3.1 Perturbative Variations of Initial Values

From the f–g representation

$$\mathbf{r} = \mathbf{r}_0 f + \mathbf{v}_0 g, \tag{4.28}$$

$$\mathbf{v} = \mathbf{r}_0 \dot{f} + \mathbf{v}_0 \dot{g}, \tag{4.29}$$

one can solve $\mathbf{r}_0$ and $\mathbf{v}_0$ in terms of t, $\mathbf{r}$ and $\mathbf{v}$. Because $f\dot{g} - \dot{f}g = 1$,

$$\mathbf{r}_0 = \mathbf{r}\dot{g} - \mathbf{v}g, \tag{4.30}$$

$$\mathbf{v}_0 = -\mathbf{r}\dot{f} + \mathbf{v}f. \tag{4.31}$$

The functions f and g (and their derivatives) must be expressed in terms of the variables at the general moment t. This can be done by changing the role of t_0 and t. Hence, the time difference and $X = \int dt/r$ change their sign. The generalized Kepler's equation takes the form

$$rX - \eta G_2(\beta, X) + \zeta G_3(\beta, X) = \tau, \tag{4.32}$$

where $\tau = t - NT$ is the time difference reduced by a number N of periods T. (This option is rarely used, but is included for completeness.) The formulae for the back-computation (from t to t_0) may thus be written as

$$\beta = 2\mu/r - \mathbf{v}^2, \tag{4.33}$$

$$\eta = \mathbf{r} \cdot \mathbf{v}, \tag{4.34}$$

$$\zeta = \mu - r\beta, \tag{4.35}$$

$$\tau = t - NT, \quad T = 2\pi\mu/(\beta\sqrt{\beta}), \quad N = \mathrm{Int}[(t + \tfrac{1}{2}T)/T], \tag{4.36}$$

$$\tau = rX - \eta G_2 + \zeta G_3, \tag{4.37}$$

$$r_0 = r - \eta G_1 + \zeta G_2, \tag{4.38}$$

$$f = 1 - \mu G_2 / r_0, \tag{4.39}$$

$$g = \tau - \mu G_3, \tag{4.40}$$

$$\dot{f} = -\mu G_1 / (r_0 r), \tag{4.41}$$

$$\dot{g} = 1 - \mu G_2 / r, \tag{4.42}$$

$$\mathbf{r}_0 = \dot{g}\mathbf{r} - g\mathbf{v}, \tag{4.43}$$

$$\mathbf{v}_0 = -\dot{f}\mathbf{r} + f\mathbf{v}. \tag{4.44}$$

These formulae must now be differentiated in the perturbative sense, line by line, to get the time derivatives of $\mathbf{r}_0^{`}$ and $\mathbf{v}_0^{`}$. The results are as follows:

$$\beta^{`} = -2\mathbf{v} \cdot \mathbf{v}^{`}, \tag{4.45}$$

$$\eta^{`} = \mathbf{r} \cdot \mathbf{v}^{`}, \tag{4.46}$$

$$\zeta^{`} = r\beta^{`}, \tag{4.47}$$

$$\tau^{`} = -\tfrac{3}{2} NT\beta^{`}/\beta, \tag{4.48}$$

and (from Kepler's equation) for $X^{`}$,

$$X^{`} = \frac{1}{r_0}\left(\tau^{`} + G_2\eta^{`} - \left(-\eta\frac{\partial G_2}{\partial \beta} + \zeta\frac{\partial G_3}{\partial \beta} - rG_3\right)\beta^{`}\right), \tag{4.49}$$

where

$$\frac{\partial G_n}{\partial \beta} = \frac{1}{2}(nG_{n+2} - XG_{n+1}). \tag{4.50}$$

Furthermore

$$G_n^{`} = G_{n-1}X^{`} + \frac{\partial G_n}{\partial \beta}\beta^{`}, \quad (n = 1, 2, 3), \tag{4.51}$$

$$r_0^{`} = -\eta^{`} G_1 + \zeta^{`} G_2 - \eta G_1^{`} + \zeta G_2^{`}, \tag{4.52}$$

$$f^{`} = -\mu G_2^{`}/r_0 + (\mu G_2/r_0^2) r_0^{`}, \tag{4.53}$$

$$g^{`} = \tau^{`} - \mu G_3^{`}, \tag{4.54}$$

$$\dot{f}^{`} = -\mu G_1^{`}/(r_0 r) + (\mu G_1/(r_0^2 r)) r_0^{`}, \tag{4.55}$$

$$\dot{g}^{`} = G_2^{`}/r, \tag{4.56}$$

$$\mathbf{r}_0^{`} = \dot{g}^{`}\mathbf{r} - g^{`}\mathbf{v} - g\mathbf{v}^{`}, \tag{4.57}$$

$$\mathbf{v}_0^{`} = -\dot{f}^{`}\mathbf{r} + f^{`}\mathbf{v} + f\mathbf{v}^{`}, \tag{4.58}$$

where $\mathbf{v}^{`} = \mathbf{F}$ is the perturbation.

These are the formulae for the computation of perturbative variations of the parametric initial conditions. If the perturbations are small, $\mathbf{r}_0$ and $\mathbf{v}_0$ change slowly, and numerical integration of the motion may be easier with the perturbative method. However, the equations of motion are rather complicated, which adds to the computational overhead.

4.3.2 Connections between Partial Derivatives

In the preceding section, the perturbative variations were presented in two forms:

1. Lagrange form

$$x_0\grave{} = -\mathbf{F} \cdot \frac{\partial \mathbf{r}}{\partial \dot{x}_0},$$

$$\dot{x}_0\grave{} = \mathbf{F} \cdot \frac{\partial \mathbf{r}}{\partial x_0}.$$

2. Poisson's form

$$x_0\grave{} = \mathbf{F} \cdot \frac{\partial x_0}{\partial \mathbf{v}},$$

$$\dot{x}_0\grave{} = \mathbf{F} \cdot \frac{\partial \dot{x}_0}{\partial \mathbf{v}}.$$

Similar expressions apply to the y and z components.

Comparing these forms, it follows that

$$\frac{\partial x_0}{\partial \mathbf{v}} = -\frac{\partial \mathbf{r}}{\partial \dot{x}_0}, \tag{4.59}$$

$$\frac{\partial \dot{x}_0}{\partial \mathbf{v}} = \frac{\partial \mathbf{r}}{\partial x_0}. \tag{4.60}$$

More results can be obtained by exchanging the moments t_0 and t (t_0 is not a special moment in any case). The connections derived may be written in matrix form:

$$\frac{\partial \mathbf{r}_0}{\partial \mathbf{v}} = -\left(\frac{\partial \mathbf{r}}{\partial \mathbf{v}_0}\right)^t \tag{4.61}$$

$$\frac{\partial \mathbf{v}_0}{\partial \mathbf{v}} = \left(\frac{\partial \mathbf{r}}{\partial \mathbf{r}_0}\right)^t \tag{4.62}$$

$$\frac{\partial \mathbf{v}}{\partial \mathbf{v}_0} = \left(\frac{\partial \mathbf{r}_0}{\partial \mathbf{r}}\right)^t \tag{4.63}$$

$$\frac{\partial \mathbf{v}_0}{\partial \mathbf{r}} = -\left(\frac{\partial \mathbf{v}}{\partial \mathbf{r}_0}\right)^t \tag{4.64}$$

(Note: Generally, for any vector $\mathbf{F}$,

$$\begin{aligned}\mathbf{F} \cdot d\mathbf{r} &= \mathbf{F} \cdot \frac{\partial \mathbf{r}}{\partial x_0} dx_0 + \mathbf{F} \cdot \frac{\partial \mathbf{r}}{\partial \dot{x}_0} d\dot{x}_0 + (y, z \text{ terms}) \\ &= \dot{x}_0\grave{}(\mathbf{F}) dx_0 - x_0\grave{}(\mathbf{F}) d\dot{x}_0 + (y, z \text{ terms}) \\ &= \mathbf{v}_0\grave{}(\mathbf{F}) \cdot d\mathbf{r}_0 - \mathbf{r}_0\grave{}(\mathbf{F}) \cdot d\mathbf{v}_0. \end{aligned} \tag{4.65}$$

This is a useful result, e.g. in orbit determination (from observations), where the condition equations can generally be written in the form $\mathbf{A} \cdot d\mathbf{r} = \Delta$ where $\mathbf{A}$ is some vector depending on the observation(s), Δ is an observational error and $d\mathbf{r}$ is the correction of $\mathbf{r}$ (due to correction of $d\mathbf{r}_0$ and $d\mathbf{v}_0$). Thus, to obtain the coefficients of the conditional equations, one may use the perturbative variation computation, replacing the perturbation $\mathbf{F}$ by $\mathbf{A}$.)

4.4 Perturbations of Orbital Elements

4.4.1 Canonical Perturbation Equations

Suppose the system is described by the Hamiltonian

$$H = K + H_1, \tag{4.66}$$

where K is the Keplerian part and $H_1 = -R$ is the perturbation. Let $(\mathbf{q}, \mathbf{p})$ be any set of canonically conjugate (generalized) coordinates and momenta. As before, the constants of motion of the Kepler problem are denoted by $\mathbf{c} = (c_1, \ldots, c_6)$. These are functions of coordinates and momenta and their rates of variation can be written

$$\dot{c}_\nu = \frac{\partial c_\nu}{\partial t} + \frac{\partial c_\nu}{\partial \mathbf{q}} \cdot \dot{\mathbf{q}} + \frac{\partial c_\nu}{\partial \mathbf{p}} \cdot \dot{\mathbf{p}} = \frac{\partial c_\nu}{\partial t} + \frac{\partial c_\nu}{\partial \mathbf{q}} \cdot \frac{\partial H}{\partial \mathbf{p}} - \frac{\partial c_\nu}{\partial \mathbf{p}} \cdot \frac{\partial H}{\partial \mathbf{q}}. \tag{4.67}$$

This is often written

$$\dot{c}_\nu = \frac{\partial c_\nu}{\partial t} + \{c_\nu, H\}, \tag{4.68}$$

where the curly brackets are called *Poisson brackets*. This bracket expression is independent of the canonical system $(\mathbf{q}, \mathbf{p})$ used (because $\{c_\nu, H\} = \dot{c}_\nu - \partial c_\nu / \partial t$). One should note that, for any function $f(\mathbf{q}, \mathbf{p}, t)$, the total time derivative is

$$\dot{f} = \frac{\partial f}{\partial t} + \{f, H\}. \tag{4.69}$$

Clearly the Poisson bracket is a linear differential operator, so that e.g. $\{f, K + H_1\} = \{f, K\} + \{f, H_1\}$. As the elements c_ν are supposed to be constants in the Keplerian motion,

$$\frac{\partial c_\nu}{\partial t} + \{c_\nu, K\} = 0. \tag{4.70}$$

Consequently

$$\dot{c}_\nu = \{c_\nu, H_1\}, \tag{4.71}$$

from which equation (4.21) follows if $H_1 = -R(\mathbf{r}, t)$ is independent of momenta. This canonical formulation is more general in the sense that H_1 may also depend on momenta; however, the earlier form (4.21) also applies when the perturbing force does not have a potential.

Equation (4.71) is independent of the canonical variables used, which is an advantage. Sometimes, however, the perturbing function is simple when written in terms of non-canonical elements. In such a case it is convenient to have the equations of motion written in terms of partial derivatives with respect to the non-canonical elements. In general it is obvious that the equations of motion then take the form

$$\dot{\mathbf{c}} = \mathbf{A}\frac{\partial H_1}{\partial \mathbf{c}}, \tag{4.72}$$

where $\mathbf{A}$ is a matrix specific to the element system $\mathbf{c}$. Simple transformation of derivatives provides the result

$$A_{ij} = \{c_i, c_j\} \tag{4.73}$$

in terms of the Poisson brackets. The next section lists the equations of motion derivable for the so-called elliptical elements of Keplerian motion.

4.4.2 Elliptical Elements

Lagrange (published in 1867–1892) derived perturbation equations for the elliptical elements. These differential equations are listed below and expressed in terms of the perturbative function R:

$$\begin{aligned}
\frac{da}{dt} &= \frac{2}{na}\frac{\partial R}{\partial M}, \qquad (n = \sqrt{\mu}/a^{3/2}),\\
\frac{de}{dt} &= \frac{-\sqrt{1-e^2}}{na^2 e}\frac{\partial R}{\partial \omega} + \frac{1-e^2}{na^2 e}\frac{\partial R}{\partial M},\\
\frac{di}{dt} &= \frac{-1}{na^2\sqrt{1-e^2}\sin(i)}\frac{\partial R}{\partial \Omega} + \frac{\cos(i)}{na^2\sqrt{1-e^2}\sin(i)}\frac{\partial R}{\partial \omega},
\end{aligned} \tag{4.74}$$

$$\frac{d\Omega}{dt} = \frac{1}{na^2\sqrt{1-e^2}\sin(i)}\frac{\partial R}{\partial i},$$

$$\frac{d\omega}{dt} = \frac{\sqrt{1-e^2}}{na^2 e}\frac{\partial R}{\partial e} - \frac{\cos(i)}{na^2\sqrt{1-e^2}\sin(i)}\frac{\partial R}{\partial i},$$

$$\frac{dM}{dt} = n - \frac{2}{na}\frac{\partial R}{\partial a} - \frac{1-e^2}{na^2 e}\frac{\partial R}{\partial e}. \tag{4.75}$$

Clearly, small eccentricity and/or inclination cause very large derivatives (infinite if the orbit is a circle and inclination zero). The reason is that, for small inclination, the angles ω and Ω can change a lot if the orbit changes a little. The same applies to M and ω, if the eccentricity is small.

In particular, for a circle, M and ω are undefined and if the inclination is zero then both ω and Ω are undefined. However, the mean longitude

$$\lambda = M + \omega + \Omega \tag{4.76}$$

is always well defined.

4.4.3 Delaunay's Variables

Delaunay (published in 1860) used for his theory of the Moon a set of variables which are generally abbreviated L, G, H, l, g and h. These *Delaunay variables* may be written in terms of the elliptic elements as:

$$L = \sqrt{\mu a}, \quad G = \sqrt{\mu a(1-e^2)}, \quad H = \sqrt{\mu a(1-e^2)}\cos(i), \tag{4.77}$$

$$\ell = M, \quad g = \omega, \quad h = \Omega. \tag{4.78}$$

These variables are *canonical* (we omit the proof). When the letter H is here reserved for one of the variables, we must use another symbol for the Hamiltonian. Often the Greek letter Φ is adopted and many authors also change the sign of the Hamiltonian, so that

$$\Phi = \frac{\mu}{2a} + R = \frac{\mu^2}{2L^2} + R. \tag{4.79}$$

The equations of motion read

$$\dot{L} = \frac{\partial \Phi}{\partial l} = \frac{\partial R}{\partial l},$$

$$\dot{G} = \frac{\partial \Phi}{\partial g} = \frac{\partial R}{\partial g},$$

$$\dot{H} = \frac{\partial \Phi}{\partial h} = \frac{\partial R}{\partial h},$$

$$\dot{l} = -\frac{\partial \Phi}{\partial L} = \frac{\mu^2}{L^3} - \frac{\partial R}{\partial L},$$
$$\dot{g} = -\frac{\partial \Phi}{\partial G} = \frac{\partial R}{\partial G},$$
$$\dot{h} = -\frac{\partial \Phi}{\partial H} = \frac{\partial R}{\partial H}. \tag{4.80}$$

One observes that all the derivatives, other than that of $l \ \ (= M)$, are of the order of magnitude of the perturbation.

4.4.4 Tremaine's Variables

A problem of considerable interest is cometary dynamics, but for this kind of problem, with high to very high eccentricities, suitable canonical elements were not available until recently (Tremaine, 2001). This new set of elements can be listed as

$$I_1,\ I_2,\ I_3,\ \ \ell,\ \theta,\ \phi, \tag{4.81}$$

where

$$I_1 = L,$$
$$I_2 = \frac{G \sin(i) \cos(g)}{\sqrt{1 - \sin^2(g) \cos^2(i)}},$$
$$I_3 = H, \tag{4.82}$$

and the new angles are defined by

$$\sin(\theta) \sin(\phi - h) = -\sin(g) \cos(i),$$
$$\sin(\theta) \cos(\phi - h) = -\cos(g),$$
$$\cos(\theta) = -\sin(g) \sin(i). \tag{4.83}$$

Here I_2 is the component of angular momentum along a line perpendicular to the z-axis and the line of apsides. The angular elements θ and ϕ are the spherical polar coordinates of the vector pointing from the origin to the aphelion of the orbit. These elements are useful for ellipses with large eccentricity. However, I_2 is indeterminate if the line of apsides coincides with the z-axis, but this problem can usually be handled by a suitable choice of the coordinate system.

Chapter 5
Numerical Integration

In this chapter we first discuss the most classical methods, the difference methods, or multi-step methods. They use previous values of derivatives to evaluate the next function value in an equidistant (in time) table. Recursive Taylor series are also considered. This latter method can usually achieve high accuracy, but may be computationally expensive.

5.1 Difference Methods

5.1.1 Operators

Consider the linear operators **A**, **B**, ..., which operate on functions $f(t)$. In what follows, linear operators commute, i.e. $\mathbf{AB} = \mathbf{BA}$. As is well known, this implies that operators have the same eigenfunctions, so that if $\mathbf{A}f(t) = \lambda_A f(t)$ then also $\mathbf{B}f(t) = \lambda_B f(t)$.

The particular operators which are most important here are:

shift	$Ef(t) = f(t+h)$,
derivative	$Df(t) = df(t)/dt$,
backward difference	$\Delta f(t) = f(t) - f(t-h)$,
central difference	$\delta f(t) = f(t+h/2) - f(t-h/2)$.

Except for the derivative, all these have the tabular interval h as a parameter. If some other interval is used, it will be specified explicitly.

All these operators commute, so that, for example,

$$D(\Delta f(t)) = D(f(t) - f(t-h)) = Df(t) - Df(t-h) = \Delta Df(t).$$

The 'deeper' reason for this is that the exponential function is an eigenfunction of these operators:

$$\begin{aligned}
Ee^{\lambda t} &= e^{\lambda h}e^{\lambda t},\\
De^{\lambda t} &= \lambda e^{\lambda t},\\
\Delta e^{\lambda t} &= (1 - e^{-\lambda h})e^{\lambda t},\\
\delta e^{\lambda t} &= (e^{\lambda h/2} - e^{-\lambda h/2})e^{\lambda t}.
\end{aligned}$$

The powers of operators are defined in the obvious way,

$$\mathbf{A}^n f(t) = \mathbf{A}(\mathbf{A}^{n-1} f(t)).$$

For example,

$$\begin{aligned}
\delta^2 f(t) &= \delta(f(t+h/2) - f(t-h/2)) = \delta f(t+h/2) - \delta f(t-h/2)\\
&= f(t+h) - 2f(t) + f(t-h)
\end{aligned}$$

is the second difference and

$$D^n f(t) = f^{(n)}(t)$$

is the nth derivative.

A function of an operator can be understood in terms of the Taylor series expansion

$$F(\mathbf{A})f(t) = \sum_{k=0}^{\infty} F^{(k)}(0)\frac{\mathbf{A}^k f(t)}{k!}. \tag{5.1}$$

For all the introduced operators, $\mathbf{A}^N f(t) = 0$ if $f(t)$ is a polynomial of degree lower than N. Thus the application of the operators and their analytical functions give reasonable results at least in the space of all polynomials.

If $f(t)$ is an eigenfunction of $\mathbf{A}$, i.e. $\mathbf{A}f(t) = \lambda_A f(t)$, then for any function F

$$F(\mathbf{A})f(t) = F(\lambda_A)f(t). \tag{5.2}$$

Finally, it is important to notice that, due to the commutative properties of the operators, the operator functions can be algebraically manipulated just like normal functions.

For example, consider the Taylor expansion

$$f(t_0 + h) = \sum_{k=0}^{\infty} \frac{h^k}{k!} D^k f(t_0). \tag{5.3}$$

This can be symbolically written

$$f(t_0 + h) = \exp(hD) f(t_0). \tag{5.4}$$

We can also write

$$\Delta f = f(t) - f(t-h) = (1 - \exp(-hD))f(t), \tag{5.5}$$

which leads to the symbolic equation

$$\Delta = 1 - \exp(-hD). \tag{5.6}$$

Solving for D gives:

$$\begin{aligned} D &= -\frac{1}{h}\ln(1-\Delta) \\ &= \frac{1}{h}\left(\Delta + \frac{\Delta^2}{2} + \frac{\Delta^3}{3} + \frac{\Delta^4}{4} + \frac{\Delta^5}{5} + \frac{\Delta^6}{6} + \frac{\Delta^7}{7} + O(\Delta)^8\right). \end{aligned} \tag{5.7}$$

This is an expression for obtaining, numerically, the value of the derivative of a tabulated function. The table below illustrates the construction of a difference table and the meaning of the different quantities:

t	f_n	Δ	Δ^2	Δ^3	Δ^3
$-3h$	f_{-3}		$\Delta^2 f_{-2}$		$\Delta^4 f_{-1}$
		Δf_{-2}		$\Delta^3 f_{-1}$	
$-2h$	f_{-2}		$\Delta^2 f_{-1}$		$\Delta^4 f_0$
		Δf_{-1}		$\Delta^3 f_0$	
$-h$	f_{-1}		$\Delta^2 f_0$		$\Delta^4 f_1$
		Δf_0		$\Delta^3 f_1$	
0	f_0		$\Delta^2 f_1$		
		Δf_1			
h	f_1				

A formula to extrapolate from previous (equidistant) tabular values $y_n,\ y_{n-1},\ y_{n-2},\ldots$ to a new value y_{n+1} can be obtained in a simple way by using the fact that $\Delta^N f_{n+1} = 0$ if $f(t)$ is a polynomial of degree less than N. Then $\Delta = 1 - E^{-1}$ and thus $\Delta^N = 1 - \binom{N}{1}E^{-1} + \binom{N}{2}E^{-2} + \cdots + (-1)^N E^{-N}$, and

$$y_{n+1} = \binom{N}{1}y_n - \binom{N}{2}y_{n-1} + \cdots - (-1)^N y_{n-(N-1)}. \tag{5.8}$$

Then, for example, extrapolation with a second-order polynomial is performed by the expression

$$y_{n+1} \approx 3y_n - 3y_{n-1} + y_{n-2}. \tag{5.9}$$

Interpolation formulae in a table can be obtained easily. It is desired to calculate the function $f(t)$ in a non-tabular point $t = t_n + \tau$, where t_n is a tabular point. Since $f(\tau) = e^{\tau D} f_n$ the relation $e^{-hD} = 1 - \Delta$ can be used to get $e^{\tau D} = (e^{hD})^{\tau/h} = (1-\Delta)^{\tau/h}$. Abbreviating $x = \tau/h$, the result is

$$
\begin{aligned}
f_{n+x} &= (1-\Delta)^{-x} f_n \\
&= 1 + \Delta f_n x + \frac{\Delta^2 f_n\, x\,(1+x)}{2} \\
&\quad + \frac{\Delta^3 f_n\, x\,(1+x)\,(2+x)}{6} \\
&\quad + \frac{\Delta^4 f_n\, x\,(1+x)\,(2+x)\,(3+x)}{24} \\
&\quad + \frac{\Delta^5 f_n\, x\,(1+x)\,(2+x)\,(3+x)\,(4+x)}{120} + \cdots . \qquad (5.10)
\end{aligned}
$$

Interpolation polynomials that fit a certain set of functional values are unique, so that this formula is just one of many possible forms in which an interpolation formula can be written. To obtain integration formulae, one may integrate the interpolation polynomial. However, methods that are more convenient for solving differential equations can be obtained more directly, as is discussed in the next section.

5.1.2 Adams and Störmer–Cowell Methods

Let a first-order differential equation be

$$\dot{y} = f(y), \qquad (5.11)$$

and let us write for the first difference

$$y_{n+1} - y_n = (\exp(hD) - 1)y_n = hF(\Delta)Dy_n = hF(\Delta)f_n, \qquad (5.12)$$

in order to compute the first difference in terms of the derivative and the differences of the derivative. Thus

$$
\begin{aligned}
F(\Delta)f_n &= \frac{\exp(hD) - 1}{hD} f_n = \frac{\Delta}{-(1-\Delta)\ln(1-\Delta)} f_n, \\
&= f_n + \frac{\Delta f_n}{2} + \frac{5\,\Delta^2 f_n}{12} + \frac{3\,\Delta^3 f_n}{8} + \frac{251\,\Delta^4 f_n}{720} \\
&\quad + \frac{95\,\Delta^5 f_n}{288} + \frac{19087\,\Delta^6 f_n}{60480} + \frac{5257\,\Delta^7 f_n}{17280} + \cdots , \qquad (5.13)
\end{aligned}
$$

which gives the explicit expression to obtain the next functional value

$$y_{n+1} = y_n + hF(\Delta)f_n. \qquad (5.14)$$

This is called the *Adams' method.*

Splitting a second-order differential equation $\ddot{y} = f(y)$ into two first-order equations $\dot{v} = f(y)$ and $\dot{y} = v$ is generally not recommended because the formulae for integration of the second-order equations usually converge much faster, thus giving more accurate results. Consider the second-order differential equation

$$\ddot{y} = f(y). \tag{5.15}$$

(It could also be $f = f(y, t)$, but not $f = f(y, \dot{y}, t)$.) Write $y_n = y(t_n)$, $t_n = nh$ where h = stepsize. The second difference is approximately proportional to the second derivative $y_{n+1} - 2y_n + y_{n-1} \approx h^2\ddot{y} = h^2 f(y_n)$. If two values y_0 and y_1 ($\approx y_0 + \dot{y}_0 h + \frac{1}{2}h^2\ddot{y}_0$) are known then new values can be computed using

$$y_{n+1} = y_n + (y_n - y_{n-1}) + h^2 f(y_n). \tag{5.16}$$

More accurate results are obtained if the value of the second difference is expressed more accurately. This can be done by fitting the second derivative f with a polynomial using the earlier known values $f_n, f_{n-1}, f_{n-2}, f_{n-3}, \ldots$ To obtain a more accurate formula for the second difference one can begin with the identity

$$\delta^2 y_n = h^2 F(\Delta) D^2 y_n, \tag{5.17}$$

and ask what the function $F(\Delta)$ is. From the differential equation one has

$$\delta^2 = h^2 F(\Delta) D^2, \tag{5.18}$$

or

$$F(\Delta) = \delta^2/(hD)^2, \tag{5.19}$$

where the operators δ and D must be expressed in terms of Δ. As

$$\delta^2 y_n = y_{n+1} - 2y_n + y_{n-1} = (\exp(hD) - 2 + \exp(-hD))y_n, \tag{5.20}$$

it is clear that

$$\delta^2 = \exp(hD) - 2 + \exp(-hD) = \frac{1}{1-\Delta} - 2 + 1 - \Delta = \frac{\Delta^2}{1-\Delta} \tag{5.21}$$

and $hD = -\ln(1 - \Delta)$. Thus

$$F(\Delta) = \frac{\Delta^2}{(1-\Delta)\ln^2(1-\Delta)} = \sum_\nu F^{(\nu)}(0)\frac{\Delta^\nu}{\nu!}. \tag{5.22}$$

When the series for F is truncated at order N, we obtain a formula which has truncation error of the order of $O(h^{N+3})$. The equation $\ddot{y} = f(y)$ can thus be integrated by

$$\begin{aligned} y_{n+1} &= 2y_n - y_{n-1} + h^2(F(0)f_n + F'(0)\Delta f_n + \tfrac{1}{2}F''(0)\Delta^2 f_n + \cdots) \\ &= 2y_n - y_{n-1} + h^2\left(f_n + \frac{\Delta^2 f_n}{12} + \frac{\Delta^3 f_n}{12} + \frac{19\,\Delta^4 f_n}{240} + \frac{3\,\Delta^5 f_n}{40} + \cdots\right). \end{aligned} \tag{5.23}$$

This formula uses only the 'old' values of the second derivative, which are already known, f_n, f_{n-1}, f_{n-2}, ..., and it is thus *explicit.* It is also often called an extrapolation formula because, in principle, it is constructed by integrating the polynomial which fits the old values of the second derivative and thus extrapolates the value over the next tabular interval $t_n \to t_{n+1}$. An implicit, and generally more accurate, equation is obtained if the next function value f_{n+1} is also used. This can be introduced into the equation by writing

$$\delta^2 y_n = h^2 \widetilde{F}(\Delta) D^2 y_{n+1} = h^2 \widetilde{F}(\Delta) D^2 \exp(hD) y_n, \tag{5.24}$$

from which

$$\widetilde{F}(\Delta) = \exp(-hD) F(\Delta) = (1 - \Delta) F(\Delta) = \frac{\Delta^2}{\ln^2(1 - \Delta)}. \tag{5.25}$$

More explicitly, one may write

$$y_{n+1} = 2y_n - y_{n-1} + h^2 f_n + \frac{h^2}{12}(f_{n+1} - 2f_n + f_{n-1}) + \cdots. \tag{5.26}$$

At first the new functional value and the differences are unknown. However, a first estimate can be obtained by summation (assuming the highest difference to remain the same, which leads to a result equivalent to that of using the extrapolation result)

$$\Delta^k f_{n+1} = \Delta^k f_n + \Delta^{k+1} f_n, \tag{5.27}$$

and, after evaluating the first estimate for y_{n+1}, one recalculates f_{n+1} and the differences. This procedure must be repeated until the result no longer changes significantly. The explicit extrapolation formula

$$y_{n+1} - 2y_n + y_{n-1} = h^2 \frac{\Delta^2}{(1 - \Delta)\ln^2(1 - \Delta)} \ddot{y}_n \tag{5.28}$$

is called *Störmer's method*, while the implicit interpolation procedure

$$y_{n+1} - 2y_n + y_{n-1} = h^2 \frac{\Delta^2}{\ln^2(1 - \Delta)} \ddot{y}_{n+1} \tag{5.29}$$

is called *Cowell's method.*

Generally, Cowell's method is more accurate and stable; however, this applies to results using the same timestep. It is often possible to obtain the same precision with Störmer's method by using a somewhat shorter timestep. In view of the necessary iterations with Cowell's method, it finally turns out that the Störmer variant is typically more efficient.

Another possibility is that of using the central differences. Express $F = \delta^2/(hD)^2$ as a function of δ^2. First solve for hD

from the relation $\delta^2 = e^{dD} - 2 + e^{-hD}$ to get $(hD)^2 = \ln^2(1 + \delta^2/2 \pm \sqrt{\delta^2 + \delta^4/4})$, and then obtain

$$\begin{aligned} F(\delta^2) &= \frac{\delta^2}{\ln^2(1 + \delta^2/2 \pm \sqrt{\delta^2 + \delta^4/4})} \\ &= 1 + \frac{\delta^2}{12} - \frac{\delta^4}{240} + \frac{31\,\delta^6}{60480} - \frac{289\,\delta^8}{3628800} + \cdots . \end{aligned} \tag{5.30}$$

This expression converges much faster than the backward difference formulae, but the method requires knowledge of future values of the second derivative, and can therefore be used only to integrate known functions.

Practical Implementation

The most practicable of the above is probably Störmer's method. Therefore, a concise discussion of its implementation is in order.

To start an integration one must construct an initial table of function values and differences. This may be done, for example, by using some other method (such as the GBS (for Gragg–Bulirsch–Stoer) extrapolation method; see later). A possible inefficiency of the starter is of no concern because it is used only in the starting procedure.

When output is needed one may want to know the velocities too. To compute them an expression of the form

$$Dy_n = \Delta y_n/h + hG(\Delta)\ddot{y}_n \tag{5.31}$$

is available. This uses the differences of function values and those of the second derivative, which both are readily available from the calculation of the coordinates. Solving for $G(\Delta)$ gives $G(\Delta) = (hD - \Delta)/(hD)^2$, which can be converted to a function of Δ using $hD = \ln(1 - \Delta)$.

One usually computes the differences Δy_n by summation and a second summation then gives the coordinates. The Störmer algorithm can thus be written

$$\Delta y_n = \Delta y_{n-1} + h^2 F(\Delta)\ddot{y}_{n-1}, \tag{5.32}$$

$$y_n = y_{n-1} + \Delta y_n, \tag{5.33}$$

$$Dy_n = \Delta y_n/h + hG(\Delta)\ddot{y}_n, \tag{5.34}$$

where the derivative expression is used only when velocities are needed. It is worth noting that this double summation formulation reduces the amount of roundoff error and can thus be recommended instead of the form (5.28). The functions F and G and their series expansions are

$$
\begin{aligned}
F(\Delta) &= \frac{\Delta^2}{(1-\Delta)\ln^2(1-\Delta)} \\
&= 1 + \Delta^2/12 + \Delta^3/12 + 19\Delta^4/240 + 3\Delta^5/40 + \cdots, \quad (5.35) \\
G(\Delta) &= -\frac{\ln(1-\Delta)+\Delta}{\ln^2(1-\Delta)} \\
&= 1/2 - \Delta/6 - \Delta^2/24 - \Delta^3/45 - 7\Delta^4/480 + \cdots. \quad (5.36)
\end{aligned}
$$

More terms can easily be obtained by use of computer algebra programs.

5.2 Runge–Kutta and Power Series Methods

5.2.1 Powers, Trigonometric and Mixed Functions

In the preceding sections, operator equalities of the form

$$
\mathbf{A} = F(\mathbf{B})\mathbf{C} \quad (5.37)
$$

have been considered, where $\mathbf{B}$ is some difference operator and $\mathbf{A}f(t)$ is the desired quantity. By expanding the function F in Taylor series, results have been obtained that are exact for polynomials $f(t)$.

Now the question arises: What happens if the function F is replaced by an interpolation polynomial? Let the polynomial P be defined by

$$
P(z_k) = F(z_k), \quad (5.38)
$$

where z_k are given complex numbers. It may be shown that the resulting difference equations are valid for the functions for which the z_k are eigenvalues of $\mathbf{B}$.

If $\mathbf{B} = \Delta$, then the condition reads $z_k = 1 - e^{\lambda h}$ and the resulting formula is accurate for the function $e^{\lambda t}$. Thus, if the formula is to be exact for the trigonometric functions $\cos(\omega t)$ and $\sin(\omega t)$, the polynomial $P(z)$ must fit the values $F(z_+)$ and $F(z_-)$, with $z_\pm = 1 - e^{i\pm\omega h} = 1 - \cos(\omega h) \pm i\sin(\omega h)$. If the polynomial also fits derivative values $F^{(k)}$, $(k = 1, 2, \ldots, n)$, at the points $z_\pm$, then the formula becomes exact for the functions $t^k e^{\pm i\omega t}$. Using real values for λ above, trigonometric functions can be replaced by exponentials. Thus it is generally possible to construct difference equations valid for powers of time, trigonometric functions, exponential functions and all products of these. The reason for this is that those are exactly the functions that satisfy linear, constant-coefficient, difference equations. For example, if a Störmer-type formula of the form

$$
\delta^2 y_n = h^2\left(1 + \frac{\Delta^2}{12} + \frac{\Delta^3}{12} + K_4\Delta^4 + K_5\Delta^5\right)\ddot{y}_n \quad (5.39)
$$

is required to be valid for the trigonometric functions $\cos(\omega t)$ and $\sin(\omega t)$, one obtains for K_4 and K_5 the expressions,

$$K_4 = -\frac{7}{6} - \frac{1}{16s^4} + \frac{\frac{37}{48} + 1/(4u^2)}{s^2} - \frac{3}{u^2} + \frac{4s^2}{u^2}, \tag{5.40}$$

$$K_5 = \frac{-24s^4 - 3u^2 + s^2(12 + 7u^2)}{24s^4u^2}, \tag{5.41}$$

where $u = \omega h$ and $s = \sin(u/2)$. These give the series expansions

$$\begin{aligned} K_4 = {} & \frac{19}{240} - \frac{863\omega^2h^2}{12096} + \frac{839\omega^4h^4}{259200} - \frac{73\omega^6h^6}{2128896} \\ & + \frac{190133\omega^8h^8}{186810624000} + \frac{4091\omega^{10}h^{10}}{448345497600} + \frac{93\omega^{12}h^{12}}{243955712000} + O(h)^{13}, \\ K_5 = {} & \frac{3}{40} + \frac{19\omega^2h^2}{6048} + \frac{83\omega^4h^4}{907200} + \frac{\omega^6h^6}{591360} \\ & - \frac{577\omega^8h^8}{93405312000} - \frac{773\omega^{10}h^{10}}{313841848320} - \frac{13633\omega^{12}h^{12}}{88921857024000} + O(h)^{13}. \end{aligned}$$

Note that the formula so obtained is valid if the solution y has the form

$$y = A\cos(\omega t) + B\sin(\omega t) + P_5(t), \tag{5.42}$$

where $P_5(t)$ is any polynomial of (at most) degree five. The results can be expected to be satisfactory if the solution can be well approximated (locally) by an expression of that form.

5.2.2 The Implicit Runge–Kutta–Gaussian (RKG) Method

Implicit methods can often provide very accurate results. The particular methods now to be discussed are based on the idea of using an interpolation polynomial with nodes in the interval $(0, h)$. The Lagrange form of such a polynomial may be written

$$\tilde{f}(t) = \sum_{k=1}^{n} L_k(t) f_k, \tag{5.43}$$

where $f_k = f(t_k)$ and L_k are polynomials satisfying $L_k(t_j) = \delta_{kj}$. In terms of the polynomial $P_k(t) = \prod_{j \neq k}(t - t_j)$, the explicit form of the L_k polynomials reads

$$L_k(t) = \frac{P_k(t)}{P_k(t_k)}. \tag{5.44}$$

If the differential equation is $\dot{y} = f(y, t)$, then the values f_k are not known in advance. However, equations for them can be formed by integrating the polynomial $\tilde{f}$:

$$y_j = y_0 + \sum_{k=1}^{n} f(y_k, t_k) \int_0^{t_j} L_k(t)dt. \tag{5.45}$$

Once the nodes t_k of the formula are known in the form $t_k = x_k h$, the integrals in the above equation can be written

$$\int_0^{t_j} L_k(t)dt = hA_{jk}, \tag{5.46}$$

where A_{jk} are constants that can be tabulated. Thus a set of non-linear equations

$$y_j = y_0 + h \sum_{k=1}^{n} A_{jk} f(y_k, t_k) \tag{5.47}$$

results, which must be solved by iteration. (The first approximations can be obtained by some simple method.) The most accurate formulae are obtained if the nodes x_k are to be the zeros of the Legendre polynomials (shifted to the interval (0,1)). As examples, the formulae based on the so-called implicit midpoint method and those based on the second-order Legendre polynomial are now listed.

One-point implicit RKG: For a first-order equation $\dot{y} = f(y)$, one adopts

$$y_1 = y_0 + hf\left(\frac{y_0 + y_1}{2}\right), \tag{5.48}$$

and for the second-order equation $\ddot{y} = f(y)$, in analogy

$$v_1 = v_0 + hf\left(\frac{y_0 + y_1}{2}\right), \tag{5.49}$$

$$y_1 = y_0 + h\frac{v_0 + v_1}{2}. \tag{5.50}$$

It can be shown that this method exactly conserves the energy of the harmonic oscillator $\ddot{y} = -\omega^2 y$ (the phase, however, is not exact).

Two-point implicit RKG: Here

$$t_1 = x_1 h, \quad t_2 = x_2 h, \tag{5.51}$$

$$y_1 = y_0 + h(a_{11} f(y_1, t_2) + a_{12} f(y_2, t_2)), \tag{5.52}$$

$$y_2 = y_0 + h(a_{21} f(y_1, t_2) + a_{22} f(y_2, t_2)), \tag{5.53}$$

$$y(h) = y_0 + h(a_{31} f(y_1, t_1) + a_{31} f(y_2, t_2)), \tag{5.54}$$

where

$$(x_1, x_2) = ((3 - \sqrt{3})/6, (3 + \sqrt{3})/6), \tag{5.55}$$

$$(a_{11}, a_{12}) = (1/4, -1/(4(3 + 2\sqrt{3}))), \tag{5.56}$$

$$(a_{21}, a_{22}) = (1/(-12 + 8\sqrt{3}), 1/4), \tag{5.57}$$

$$(a_{31}, a_{32}) = (1/2, 1/2). \tag{5.58}$$

Similar methods can be obtained for the integration of a second-order differential equation. If the equation has the form $\ddot{y} = f(y, t)$, the polynomial approximation for f is integrated twice to get

$$t_1 = x_1 h, \quad t_2 = x_2 h, \tag{5.59}$$

$$y_1 = y_0 + \dot{y}_0 t_1 + h^2 (b_{11} f(y_1, t_2) + b_{12} f(y_2, t_2)), \tag{5.60}$$

$$y_2 = y_0 + \dot{y}_0 t_2 + h^2 (b_{21} f(y_1, t_2) + b_{22} f(y_2, t_2)), \tag{5.61}$$

$$y(h) = y_0 + \dot{y}_0 h + h^2 (b_{31} f(y_1, t_1) + b_{31} f(y_2, t_2)), \tag{5.62}$$

$$\dot{y}(h) = \dot{y}_0 + h(a_{31} f(y_1, t_1) + a_{32} f(y_2, t_2)), \tag{5.63}$$

where

$$(b_{11}, b_{12}) = \left(\frac{1}{36}, \frac{(-3 + \sqrt{3})}{36(3 + 2\sqrt{3})} \right), \tag{5.64}$$

$$(b_{21}, b_{22}) = \left(\frac{(3 + \sqrt{3})}{36(-3 + 2\sqrt{3})}, \frac{1}{36} \right), \tag{5.65}$$

$$(b_{31}, b_{32}) = \left(\frac{(-1 + \sqrt{3})}{(4(-3 + 2\sqrt{3}))}, \frac{(1 + \sqrt{3})}{(4(3 + 2\sqrt{3}))} \right). \tag{5.66}$$

These simple methods have an $O(h^4)$ truncation error. But with n nodes the errors are generally of order $O(h^{2n})$ and thus very accurate methods can be constructed. However, the necessary iterations often make these methods rather expensive in computational overhead.

5.2.3 Taylor Series Solution of N-Body Equations

Steffensen's Method

One could derive the Taylor series coefficients for the solution of a differential equation by repeated differentiation. In general this, however, quickly leads to very complicated expressions. In that case it is often simpler to derive recursive formulae for the power series coefficients. This procedure is known as *Steffensen's method* (Steffensen, 1907). Uniquely defined rules are not employed here, so that the method can best be explained by examples.

Non-Linear Oscillator

The equation is

$$\ddot{y} = -y + \epsilon y^2. \tag{5.67}$$

Let

$$y = \sum_{k=0}^{\infty} Y_k t^k. \tag{5.68}$$

For y^2 one gets

$$y^2 = \sum_{k=0}^{\infty} \left(\sum_{l=0}^{k} Y_l Y_{k-l} \right) t^k \tag{5.69}$$

and, for the second derivative,

$$\ddot{y} = \sum_{k=0}^{\infty} k(k-1)\, Y_k t^{k-2} = \sum_{k=0}^{\infty} (k+2)(k+1)\, Y_{k+2} t^k. \tag{5.70}$$

Substituting into the differential equation gives

$$\sum_{k=0}^{\infty} (k+2)(k+1)\, Y_{k+2} t^k = -\sum_{k=0}^{\infty} Y_k t^k + \epsilon \sum_{k=0}^{\infty} \left(\sum_{l=0}^{k} Y_l Y_{k-l} \right) t^k. \tag{5.71}$$

Comparison of coefficients of the same power of t gives the recursion formula

$$Y_{k+2} = \frac{1}{(k+2)(k+1)} \left(-Y_k + \epsilon \left(\sum_{l=0}^{k} Y_l Y_{k-l} \right) \right). \tag{5.72}$$

This permits all the coefficients to be evaluated, beginning with $Y_0 = y(0)$ and $Y_1 = \dot{y}_0$. For this example,

$$\begin{aligned} Y_2 &= \frac{1}{2}(-Y_0 + \epsilon\, Y_0^2), \\ Y_3 &= \frac{1}{6}(-Y_1 + \epsilon(Y_0 Y_1 + Y_1 Y_0)), \\ Y_4 &= \frac{1}{12}(-Y_2 + \epsilon(Y_0 Y_2 + Y_1 Y_1 + Y_2 Y_0)), \end{aligned} \tag{5.73}$$

and so on.

Two-Body Orbit

Let us take the equation of motion in the form

$$\ddot{\mathbf{r}} = -\mathbf{r}/r^3 \tag{5.74}$$

and introduce the auxiliary variables $I = r^2$ and $\Omega = 1/r^3 = I^{-1.5}$. By differentiation of the relation $I^{1.5}\Omega = 1$, one obtains

$$1.5\dot{I}\Omega + I\dot{\Omega} = 0. \tag{5.75}$$

By substituting series expansions

$$\begin{aligned} \mathbf{r} &= \sum \mathbf{r}_k t^k, & \ddot{\mathbf{r}} &= \sum (k+2)(k+1)\mathbf{r}_{k+2} t^k, \\ I &= \sum I_k t^k, & \dot{I} &= \sum (k+1) I_{k+1} t^k, \\ \Omega &= \sum \Omega_k t^k, & \dot{\Omega} &= \sum (k+1)\Omega_{k+1} t^k, \end{aligned} \tag{5.76}$$

one gets from the relation $I = \mathbf{r}^2$ the equation

$$I_k = \sum_{l=0}^{k} \mathbf{r}_l \cdot \mathbf{r}_{k-l} \tag{5.77}$$

and the differential equation $1.5\dot{I}\Omega + I\dot{\Omega} = 0$ gives

$$\left(1.5\sum (k+1) I_{k+1} t^k\right)\left(\sum \Omega_k t^k\right) + \left(\sum I_k t^k\right)\left(\sum (k+1)\Omega_{k+1} t^k\right) = 0. \tag{5.78}$$

Carrying out the series multiplications, we obtain the equation

$$1.5\sum_{l=0}^{k} (l+1) I_{l+1}\Omega_{k-l} + \sum_{l=0}^{k} (k+1-l)\Omega_{k+1-l} I_l = 0, \tag{5.79}$$

from which it is possible to solve for the highest-index Ω coefficient

$$\Omega_{k+1} = -\frac{1}{I_0(k+1)}\left(1.5(k+1)I_{k+1}\Omega_0 + \sum_{l=0}^{k-1} (k + l/2 + 3/2)\Omega_{k-l} I_{l+1}\right). \tag{5.80}$$

In addition the relation $\ddot{\mathbf{r}} = -\Omega\mathbf{r}$ is required, which, in terms of a series, is

$$\sum (k+2)(k+1)\mathbf{r}_{k+2} t^k = -\sum \left(\sum_{l=0}^{k} \Omega_l \mathbf{r}_{k-l}\right) t^k. \tag{5.81}$$

This implies for the $\mathbf{r}$ series the recursive formula

$$\mathbf{r}_{k+2} = -\frac{1}{(k+2)(k+1)} \sum_{l=0}^{k} \Omega_l \mathbf{r}_{k-l}. \tag{5.82}$$

After some modification of index notation these results can be written in the form of the following algorithm. We know $\mathbf{r}_0$ and $\mathbf{r}_1 = \dot{\mathbf{r}}_0$, from which we compute $I_0 = \mathbf{r}_0^2$, $\Omega_0 = I_0^{-1.5}$, and generally

$$
\begin{aligned}
I_n &= \sum_{l=0}^{n} \mathbf{r}_l \cdot \mathbf{r}_{n-l}, \\
\Omega_n &= -\frac{1}{nI_0}\left(1.5nI_n\Omega_0 + \sum_{k=0}^{n-2}\left(n+\frac{1}{2}+\frac{k}{2}\right)\Omega_{n-1-k}I_{k+1}\right), \\
\mathbf{r}_{n+1} &= -\frac{1}{n(n+1)}\sum_{l=0}^{n-1}\Omega_l\mathbf{r}_{n-1-l}.
\end{aligned} \tag{5.83}
$$

N-Body Taylor Algorithm

The case of N bodies is not much more complicated, in principle. The essential difference comes from summation over interactions. Also, the number of auxiliary variables is large. The equations are

$$
\ddot{\mathbf{r}}_j = -\sum_{i \neq j} m_i \frac{\mathbf{r}_j - \mathbf{r}_i}{r_{ij}^3}. \tag{5.84}
$$

Let $\Omega^{ij} = 1/r_{ij}^3$, $\mathbf{x}^{ij} = \mathbf{r}_j - \mathbf{r}_i$ and $I^{ij} = |\mathbf{x}^{ij}|^2$, where superscripts are used in the auxiliary variables to avoid confusion with series indices. The equations into which the series are to be substituted are

$$
\begin{aligned}
\ddot{\mathbf{r}}_j &= -\sum_{i \neq j} m_i \Omega^{ij}\mathbf{x}^{ij}, \\
\mathbf{x}^{ij} &= \mathbf{r}_j - \mathbf{r}_i, \\
I^{ij} &= |\mathbf{x}^{ij}|^2, \\
1.5\dot{I}^{ij}\Omega^{ij} + I^{ij}\dot{\Omega}^{ij} &= 0.
\end{aligned} \tag{5.85}
$$

When the series

$$
\begin{aligned}
\mathbf{r}^j &= \sum \mathbf{r}_k^j t^k, \\
\mathbf{x}^{ij} &= \sum \mathbf{x}_k^{ij} t^k, \\
I^{ij} &= \sum I_k^{ij} t^k, \\
\Omega^{ij} &= \sum \Omega_k^{ij} t^k
\end{aligned} \tag{5.86}
$$

are substituted, in analogy with the two-body case, one obtains the recursions

$$
\begin{aligned}
\mathbf{x}_n^{ij} &= \mathbf{r}_n^j - \mathbf{r}_n^i, \qquad I_n^{ij} = \sum_{l=0}^{n} \mathbf{x}_l^{ij} \cdot \mathbf{x}_{n-l}^{ij}, \\
\Omega_n^{ij} &= -\frac{1}{nI_0^{ij}}\left(1.5nI_n^{ij}\Omega_0^{ij} + \sum_{k=0}^{n-2}\left(n+\frac{1}{2}+\frac{k}{2}\right)\Omega_{n-1-k}^{ij}I_{k+1}^{ij}\right), \\
\mathbf{r}_{n+1}^j &= -\sum_{i \neq j} m_i \frac{1}{n(n+1)}\sum_{l=0}^{n-1}\Omega_l^{ij}\mathbf{x}_{n-1-l}^{ij},
\end{aligned} \tag{5.87}
$$

where subscripts are series indices and superscripts are variable name indices.

5.2.4 Radius of Convergence in a Kepler Orbit

When using power series it is important to know, approximately at least, the radius of convergence.

For two-body motion, it is possible to determine this radius analytically. From the theory of functions it is known that radius of convergence is the distance of the nearest singularity in the complex time plane. From the equation $\ddot{\mathbf{r}} = -\mathbf{r}/r^3$ we guess that the singularity is at $r = 0$. From the elliptic equations

$$r = a(1 - e\cos(E)), \quad E - e\sin(E) = M, \tag{5.88}$$

the value of M at which $r = 0$ can be computed, from which the radius of convergence follows:

$$\cos(E) = 1/e, \tag{5.89}$$

$$\sin(E) = \sqrt{1 - 1/e^2} = \frac{i}{e}\sqrt{1 - e^2}, \tag{5.90}$$

and thus

$$E = \text{atan}\left(i\sqrt{1 - e^2}\right) = i\ln\left(\frac{1 + \sqrt{1 - e^2}}{e}\right). \tag{5.91}$$

Therefore the singularity is at

$$M_s = i\left(\ln\left(\frac{1 + \sqrt{1 - e^2}}{e}\right) - \sqrt{1 - e^2}\right). \tag{5.92}$$

When $e \to 0$ then $|M_s| \sim \ln(1/e) \to \infty$. If the eccentricity is large ($e \sim 1$), then $\beta = \sqrt{1 - e^2}$ is small and

$$M_s = i\left(\frac{1}{3}\beta^3 + \frac{1}{5}\beta^5 + \frac{1}{7}\beta^7 + \cdots\right). \tag{5.93}$$

From periodicity, singularities also occur at

$$M_{sk} = \pm M_s + 2\pi k. \tag{5.94}$$

If the mean anomaly has the value M_0, which satisfies $|M_0| < \pi$ (otherwise reduce by removing multiples of 2π), the distance of M_s from M_0 provides the M series radius of convergence:

$$R_M = \sqrt{M_0^2 + M_s^2}. \tag{5.95}$$

At pericentre, R_M is shortest $= |M_s|$. In the case of large eccentricity this value is

$$R_{min} \approx \tfrac{1}{3}(1 - e^2)^{1.5}. \tag{5.96}$$

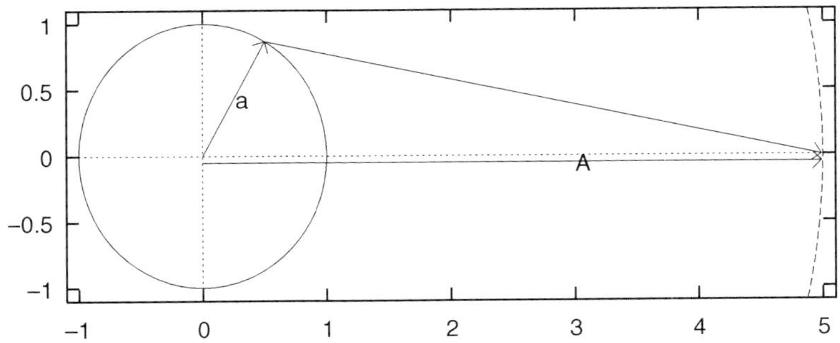

Fig. 5.1. Estimating the radius of convergence in a planetary system.

When measured in time, the radius of convergence is

$$R_t = R_M/\dot{M} = R_M a\sqrt{a/\mu} \underbrace{\approx \tfrac{1}{3}p\sqrt{p/\mu}}_{\text{if } e\sim 1}. \tag{5.97}$$

5.2.5 Radius of Convergence in a Planetary System

Exact computation of the radius of convergence (R_C) is not possible in a many-body system. However, if a planar system of planets of very small mass is considered, by neglecting perturbations (and so producing a zeroth-order estimate) R_C may be estimated. The longitude difference, the angle between a and A in Figure 5.1, is called θ in the text. It is a linear function of time, $\theta = (n_1 - n_2)t$, where n_1 and n_2 are the mean motions. It is possible to determine R_C for θ series (R_θ) and obtain $R_t = R_\theta/\dot{\theta}$. From Figure 5.1 one sees

$$\Delta^2 = a^2 + A^2 - 2aA\cos(\theta), \tag{5.98}$$

and the singularity condition $\Delta = 0$ gives

$$\cos(\theta) = \frac{1}{2}\left(\frac{a}{A} + \frac{A}{a}\right). \tag{5.99}$$

According to Euler's equation $\cos(\theta) = \frac{1}{2}(\exp(i\theta) + 1/\exp(i\theta))$ and comparison with equation (5.99) gives $\exp(i\theta) = a/A$ or

$$\theta_s = i\ln\left(\frac{A}{a}\right), \tag{5.100}$$

and generally (if $|\theta| < \pi$, otherwise reduce)

$$R_\theta = \sqrt{\theta^2 + \ln^2\left(\frac{A}{a}\right)}. \tag{5.101}$$

Using complex numbers this can be put into an easily remembered form. Let the coordinates of planet 1 be x_1, y_1 and x_2, y_2 of planet 2. Then $x_1 + iy_1 = a\exp(i\theta_1)$ and $x_2 + iy_2 = A\exp(i\theta_2)$. Thus their ratio is

$$\frac{(x_1 + iy_1)}{(x_2 + iy_2)} = \frac{a}{A}\exp(i(\theta_1 - \theta_2)). \tag{5.102}$$

Then

$$\ln\left(\frac{x_1 + iy_1}{x_2 + iy_2}\right) = \ln\left(\frac{a}{A}\right) + i\underbrace{(\theta_1 - \theta_2)}_{=\theta}. \tag{5.103}$$

The square of the absolute value of this is $= \ln^2(A/a) + \theta^2$, and consequently

$$R_\theta = \left|\ln\left(\frac{x_1 + iy_1}{x_2 + iy_2}\right)\right|. \tag{5.104}$$

It is clear that *the radius of convergence does not depend on the perturbing masses*. Thus even a very small perturbation can make it necessary to use a significantly shorter timestep.

If the orbits in a planetary system are nearly circular, the radius of convergence can be roughly estimated by taking the minimum over the elliptic orbits and perturbation terms. In practice, the useful stepsize can usually only be a small fraction of R, and it seems that no method can exceed R.

5.2.6 Radius of Convergence in a Hot Star Cluster

For a system of N bodies, in which the relative speeds are large (thus relative motions are roughly rectilinear in a short time), the radius of convergence can be estimated as follows. Consider the relative position vector $\mathbf{r}$ of two stars. Assume $\dot{\mathbf{r}} =$ constant. Then $r^2 = r_0^2 + 2\mathbf{r}_0 \cdot \mathbf{v}_0 t + v_0^2 t^2 = 0$ defines the radius of convergence. Write $\mathbf{r}_0 \cdot \mathbf{v}_0 = r_0 v_0 \cos(\theta)$ and $t = (r_0/v_0)\tau$ so that

$$r^2 = r_0^2(1 + 2\tau\cos(\theta) + \tau^2) = 0, \tag{5.105}$$

which gives

$$\tau = -\cos(\theta) \pm \sqrt{\cos^2(\theta) - 1} = -(\cos(\theta) \pm i\sin(\theta)), \tag{5.106}$$

or

$$t = -\frac{r_0}{v_0}\exp(\pm i\theta). \tag{5.107}$$

The radius of convergence for the force term due to this interaction is thus

$$R_t = \left|-\frac{r_0}{v_0}\exp(\pm i\theta)\right| = \frac{r_0}{v_0}. \tag{5.108}$$

Consequently an estimate for the entire system is

$$R_t \approx \min_{i,j} \frac{|\mathbf{r}_i - \mathbf{r}_j|}{|\dot{\mathbf{r}}_i - \dot{\mathbf{r}}_j|}. \tag{5.109}$$

5.3 Extrapolation Methods and Substep Algorithms

A popular contemporary numerical integration method is the Gragg–Bulirsch–Stoer (GBS; Gragg, 1964, 1965; Bulirsch and Stoer, 1966; Press et al., 1986) extrapolation method. This is a self-starting one-step method which uses a simple basic method with different stepsizes and then extrapolates the result to zero steplength using a polynomial (Gragg) or a rational function (Bulirsch–Stoer).

The principle of extrapolation methods is to use a simple basic integrator which is used, with shorter and shorter stepsize, to compute the solution of a differential equation over a fixed interval of time (or, more generally, an interval of the independent variable). The results are then extrapolated to zero steplength using a rational function or, alternatively, a polynomial.

5.3.1 First-Order Differential Equations

Midpoint Method

Let Δ be a stepsize and $h = \Delta/n$ a substepsize. With the differential equation $\dot{y} = f(y)$ one defines

$$\begin{aligned}
z_0 &= y(t_0), \\
z_1 &= z_0 + hf(t_0, y_0), \\
&\vdots \\
z_{m+1} &= z_{m-1} + 2hf(z_m), \qquad m = 1, 2, \ldots, n-1, \\
y(t_0 + \Delta) &\approx y_n = \frac{1}{2}(z_n + z_{n-1} + hf(z_n))
\end{aligned}$$

(modified midpoint method). (5.110)

Bulirsch and Stoer (1966) showed that, with a fixed Δ, the error of this method is an even function of h:

$$y_n = y(t_0 + \Delta) + \alpha_2 h^2 + \alpha_4 h^4 + \alpha_6 h^6 + \cdots . \tag{5.111}$$

Thus, after having done the integration with several values of the substepsize h, one is able to extrapolate the result efficiently to zero stepsize. This is the BS1 method. Details of the extrapolation can be found e.g. in Press et al. (1986).

5.3.2 Second-Order Differential Equations

For a second-order differential equation $\ddot{y} = g(y)$ one may use the method known as 'leapfrog', which usually gives an algorithm that is significantly more efficient than the splitting of the second-order equation into first-order ones.

In the leapfrog, with coordinates y and velocities $v = \dot{y}$, one starts with initial values y_0 and v_0 as

$$y_{0.5} = y_0 + \frac{h}{2}v_0, \quad v_1 = v_0 + hg(y_{0.5}), \tag{5.112}$$

and continues with

$$y_{m-0.5} = y_{m-1.5} + hv_{m-1}, \quad v_m = v_{m-1} + hg(y_{m-0.5}), \quad m = 2, \ldots, n, \tag{5.113}$$

so that the final coordinate value is given by

$$y_n = y_{n-0.5} + \frac{h}{2}v_n. \tag{5.114}$$

The error of this algorithm has a similar expansion as that for the modified midpoint method. As in the BS1 method one extrapolates the results v_n and y_n to zero steplength, thus producing results of the BS2 algorithm.

5.3.3 Composite Methods

Consider a differential equation

$$\dot{\mathbf{y}} = \mathbf{A}(\mathbf{y}) + \mathbf{B}(\mathbf{y}), \tag{5.115}$$

where the functions $\mathbf{A}$ and $\mathbf{B}$ are such that the equations

$$\dot{\mathbf{y}} = \mathbf{A}(\mathbf{y}), \tag{5.116}$$

$$\dot{\mathbf{y}} = \mathbf{B}(\mathbf{y}) \tag{5.117}$$

are individually easily integrable. The solution for the total system (5.115) may be approximated by the composition of the solutions of the two equations. If we denote by $A(h)$ and $B(h)$ the operation of advancing the vector $\mathbf{y}$ over a timestep of length h using the differential equations $\dot{\mathbf{y}} = \mathbf{A}(\mathbf{y})$ or $\dot{\mathbf{y}} = \mathbf{B}(\mathbf{y})$, we can write one step in a simple leapfrog-like algorithm in the symbolic form

$$A(h/2)B(h)A(h/2). \tag{5.118}$$

Here we first use the A equation over a timestep of length $h/2$ and adopt the result ($\mathbf{y}$) as initial value for further advancement by the B equation over the timestep h, followed by the second (half-step) A-type motion. Repeating the operation, one can obtain approximations for the solution of the differential equation (5.115) over a longer interval ($= nh$). This can be written as

$$A(h/2)(B(h)A(h))^{n-1}B(h)A(h/2). \tag{5.119}$$

As one can see, this algorithm is time-reversible and the error is quadratic, i.e.

$$\delta y = \alpha_2 h^2 + \alpha_4 h^4 + \cdots , \tag{5.120}$$

so that using the GBS extrapolation method is efficient.

5.3.4 The Leapfrog for Second-Order Equations

Let the differential equation be

$$\ddot{\mathbf{r}} = \mathbf{F}(\mathbf{r}). \tag{5.121}$$

Denoting the velocity of the point $\dot{\mathbf{r}}$ by $\mathbf{v}$, then the motion, over a short timestep h, can be approximated by first moving the point $\mathbf{r}$ linearly over a half-step $h/2$ by

$$\mathbf{r}_{1/2} = \mathbf{r}_0 + \mathbf{v}_0 h/2. \tag{5.122}$$

Then take into account the acting acceleration $\mathbf{F}$ by changing the velocity according to

$$\mathbf{v}_1 = \mathbf{v}_0 + h\mathbf{F}(\mathbf{r}_{1/2}) \tag{5.123}$$

and proceed by moving the position again to

$$\mathbf{r}_1 = \mathbf{r}_{1/2} + \mathbf{v}_1 h/2. \tag{5.124}$$

In this way the approximate position and velocity at time $t = t_0 + h$ have been computed. Repeating the sequence of operations, the coordinates and velocities at later times are obtained in a stepwise process.

More generally, let $\mathbf{r}_0 = \mathbf{r}(0)$, $\mathbf{v}_0 = \mathbf{v}(0)$ be the initial values at $t = 0$. Then the algorithm to approximate the motion over a large number of steps up to the time $t_n = nh$ becomes

$$\mathbf{r}_{1/2} = \mathbf{r}_0 + \mathbf{v}_0 h/2, \tag{5.125}$$

$$\mathbf{v}_1 = \mathbf{v}_0 + h\mathbf{F}(\mathbf{r}_{1/2}), \tag{5.126}$$

$$\mathbf{r}_{3/2} = \mathbf{r}_{1/2} + \mathbf{v}_1 h, \tag{5.127}$$

$$\mathbf{v}_2 = \mathbf{v}_1 + h\mathbf{F}(\mathbf{r}_{3/2}), \tag{5.128}$$

$$\vdots \tag{5.129}$$

$$\mathbf{r}_{n-1/2} = \mathbf{r}_{n-3/2} + h\mathbf{v}_{n-1}, \tag{5.130}$$

$$\mathbf{v}_n = \mathbf{v}_{n-1} + h\mathbf{F}(\mathbf{r}_{n-1/2}), \tag{5.131}$$

$$\mathbf{r}_n = \mathbf{r}_{n-1/2} + \mathbf{v}_n h/2. \tag{5.132}$$

Thus the half-steps appear only in the beginning and at the end of the computation because the linear motion at the end of the previous elementary leapfrog step can be concatenated with the starting half-step of the next sequence of operations. The method is of the second order in a long integration (one-step error $O(h^3)$), i.e. if integrated over a long interval of time with shorter and shorter steps, the error decreases as $O(h^2)$. The process is also time-reversible (if roundoff error can be neglected), which makes it possible to apply the GBS extrapolation method.

5.3.5 Alternative for Midpoint Formulation

One may construct the substep algorithm for a first-order equation in a way that makes it look like leapfrog.

Let the equation be

$$\dot{z} = f(z). \tag{5.133}$$

It is now possible to split this into two equations:

$$\dot{x} = f(y), \quad \dot{y} = f(x). \tag{5.134}$$

One leapfrog-like integration step can now be written as

$$x_{1/2} = x_0 + \frac{h}{2}f(y_0), \tag{5.135}$$

$$y_1 = y_0 + hf(x_{1/2}), \tag{5.136}$$

$$x_1 = x_{1/2} + \frac{h}{2}f(y_1). \tag{5.137}$$

This can naturally be augmented to many steps just like the leapfrog above. An advantage of this is that one may write a single algorithm to be used for both first- and second-order equations.

Figure 5.2 illustrates the effect of leapfrog stepsize when applied to a simple elliptic motion. In this experiment the starting distance was = 1 and the velocity, perpendicular to the starting position vector, was = 1.1. When the mass was = 1 the eccentricity of the orbit becomes = 0.21. The thick line represents the results with a

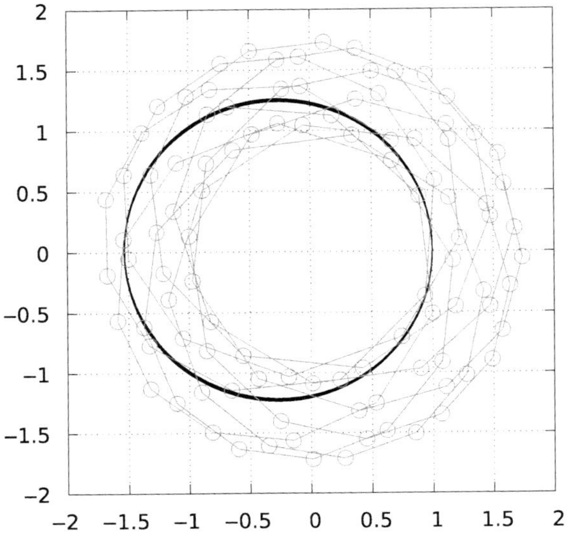

Fig. 5.2. Elliptic motion computed with leapfrog. The eccentricity was set to $e = 0.21$.

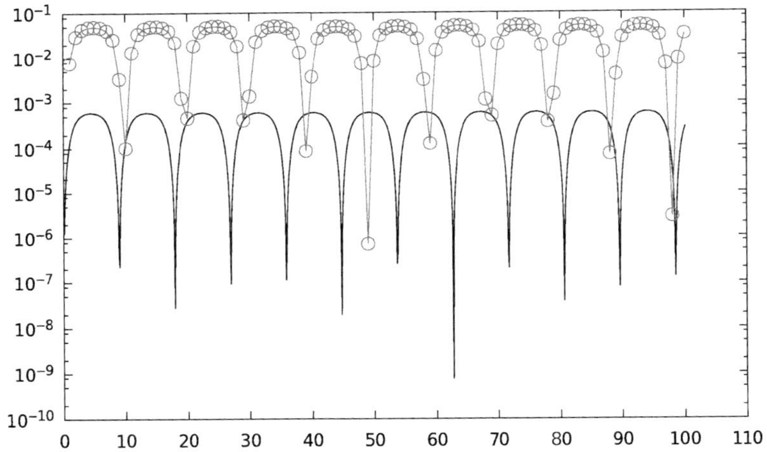

Fig. 5.3. Relative energy errors in leapfrog integrations of elliptic motions. The used stepsizes were $h = 0.1$ and $h = 1$.

stepsize $h = 0.1$ while the clearly erroneous result was obtained with $h = 1$. Figure 5.3 gives the obtained (relative) energy errors in the integrations of the orbits of Figure 5.2. It is obvious that the error behaviour is regular even when the stepsize is so large ($h = 1$ in this experiment) that the orbit in Figure 5.2 does not look like an ellipse any more. The reason is that the leapfrog method actually has a Hamiltonian that remains constant in the integration. This, so-called surrogate Hamiltonian, is not the true one, but can be expressed as a power series in terms of the stepsize. That quantity is, however, too complicated to be of much use.

5.3.6 Accuracy Checks for Extrapolation

In extrapolation methods, e.g. in Bulirsch and Stoer (1966), the integration errors can be estimated as the difference of the last two extrapolation results when stepsizes are shortened. The results are usually also used to estimate a reasonable length for the next integration step.

Sometimes it has been required that the relative error (dx/x) of every quantity (x) is smaller than a tolerance (ϵ), i.e. $|dx| < \epsilon|x|$. However, in some cases this does not work because the quantity x can be very small, even $= 0$. Thus it is better to try to estimate some natural magnitude of the quantity and then one may compare the error estimate with that. For coordinates x, y, z of a vector $\mathbf{r} = (x, y, z)$, a reasonable estimate is $|dx| < \epsilon|\mathbf{r}|$. For many quantities sometimes more complicated estimates may be needed.

Occasionally it happens that the stepsize becomes very small; then there is the possibility that the stepsize is too short. Because the extrapolation algorithms use the derivative differences in the extrapolation operation, a too short step causes the differences to be affected considerably by the roundoff. A consequence can be that the stepsize cannot increase any more and the computation stops in practice. This is a rare phenomenon, but happens sometimes. A possible cure is that one tests a longer (maybe much longer) step.

The value of the tolerance ϵ is what one must estimate when writing a code. In N-body computations a reasonable value often is something like 10^{-12}, but if such an accuracy is not needed larger values may be used. In general the accuracy that one can achieve is two or three figures less than the computer's accuracy.

Other possibilities are discussed in Section 7.4.6.

5.3.7 Variational Equations

If the chosen integration method is simple, for example leapfrog, it is easy to compute the variational equation solution.

Let

$$\ddot{x} = f(x) \tag{5.138}$$

be the equation. Then the variational equation is

$$\ddot{\delta x} = \frac{\partial f(x)}{\partial x}\delta x, \tag{5.139}$$

or one may write

$$\dot{v} = f(x), \qquad \dot{x} = v \tag{5.140}$$

$$\dot{\delta v} = \frac{\partial f(x)}{\partial x}\delta x, \quad \dot{\delta x} = \delta v. \tag{5.141}$$

The simple way of solving the variational equations is to first write a code for solving the quantities x and v and then one differentiates the code line by line. For the leapfrog the equations are then

$$x_{1/2} = x_0 + \frac{h}{2} v_0, \tag{5.142}$$

$$\delta x_{1/2} = \delta x_0 + \frac{h}{2} \delta v_0, \tag{5.143}$$

$$v_1 = v_0 + h f(x_{1/2}), \tag{5.144}$$

$$\delta v_1 = \delta v_0 + h \frac{\partial f(x_{1/2})}{\partial x_{1/2}} \delta x_{1/2}, \tag{5.145}$$

$$x_1 = x_{1/2} + \frac{h}{2} v_1, \tag{5.146}$$

$$\delta x_1 = \delta x_{1/2} + \frac{h}{2} \delta v_1. \tag{5.147}$$

In more complicated systems the operation can be done in essentially the same way: just differentiating the algorithm line by line. This gives the exact solution for the variation of the algorithm and that is likely to be the best one can do.

The extrapolation algorithms can be used in analogous ways to the coordinates, velocities and the variations.

From the values obtained for the variations δx and δv one can determine the Lyapunov exponents:

$$\lambda = \lim_{t \to \infty} \ln(|d\mathbf{x}|)/t,$$

and the e-folding times of variations $= 1/\lambda$, which are important in obtaining the rate of chaos in the system. These are also explained in Section 6.2.5.

5.4 High-Order Leapfrogs

Use the simplified notation $\mathbf{X} = (\mathbf{q}, \ldots)$ and $\mathbf{V} = (\mathbf{p}, \ldots)$, where the dots mean possible other variables that are needed. We can simply write a single leapfrog step in the form

$$\begin{aligned} \mathbf{X}_{1/2} &= \mathbf{X}_0 + \tfrac{1}{2} h \mathbf{X}'(\mathbf{V}_0), \\ \mathbf{V}_1 &= \mathbf{V}_0 + h \mathbf{V}'(\mathbf{X}_{1/2}), \\ \mathbf{X}_1 &= \mathbf{X}_{1/2} + \tfrac{1}{2} h \mathbf{X}'(\mathbf{V}_1), \end{aligned} \tag{5.148}$$

where $\mathbf{X}'$ and $\mathbf{V}'$ are the derivative vectors for $\mathbf{X}$ and $\mathbf{V}$, h is the stepsize and indices 0, 1/2 and 1 mean correspondingly the values in the beginning, midpoint and end of the step.

Writing $\mathbf{L}(h)$ for a leapfrog like (5.148), the higher-order leapfrog methods (Yoshida, 1990) (Y90) may be written

$$\begin{aligned}
&\mathbf{L}_{Y90}(h) \\
&= \prod_{k=1}^{N} \mathbf{L}(d_k h) \\
&= \prod_{k=1}^{N} [\mathbf{X}(d_k h/2)\mathbf{V}(d_k h)\mathbf{X}(d_k h/2)] \\
&= \mathbf{X}(d_1 h/2)\left[\prod_{k=1}^{N-1} \mathbf{V}(d_k h)\mathbf{X}((d_k + d_{k+1})h/2)\right]\mathbf{V}(d_N h)\mathbf{X}(d_N h/2).
\end{aligned} \tag{5.149}$$

The number of function evaluations N, the one-step error (order) err_1 and the substep lengths d_k are as follows:

1. In the fourth-order method we have

$$N = 3, \quad err_1 = O(h^5)$$

and the values of the parameters d_k are

$$\{d_1, d_2, d_3\} = \left\{\frac{1}{2 - 2^{1/3}}, \frac{-2^{1/3}}{2 - 2^{1/3}}, \frac{1}{2 - 2^{1/3}}\right\}.$$

2. In the sixth-order we have

$$N = 7, \quad err_1 = O(h^7)$$

and the d_k as given in Yoshida (1990) read

$$\begin{aligned}
&\{d_1, d_2, d_3, d_4, d_5, d_6, d_7\} \\
&= \{0.7845136104775599, 0.2355732133593569, \\
&\quad - 1.17767998417887, 1.315186320683906, \\
&\quad - 1.17767998417887, 0.2355732133593569, \\
&\quad\ 0.7845136104775599\}.
\end{aligned}$$

Chapter 6
Symplectic Integration

Symplectic integration is used especially in studying motions in planetary systems. Therefore it may be useful to become accustomed to the most used coordinate systems in those investigations. In Figure 6.1 the heliocentric coordinates are illustrated in a simple way. Similarly Figure 6.2 explains the Jacobian coordinates.

Motion in a Hamiltonian system is equivalent to a canonical transformation, so that it is desirable to have numerical methods in which each integration step is a canonical transformation. Such methods have become popular in recent years and they are generally called 'symplectic methods'.

The idea of symplectic integration is thus to consider the transformation of the coordinates and momenta at some time t to a later time $t + \tau$ to be a canonical transformation. The symplectic method then uses a series of canonical transformations to propagate the system forward in time. As with any numerical scheme, errors are induced by the nature of the symplectic propagation used, but in all cases this can be expressed by an error Hamiltonian.

It has turned out that symplectic integrators usually produce more reliable results, in long computations, than the classical integration methods. This is indeed normally the case, although symplectic integrators may have fairly large local errors. Moreover, it is known that symplectic methods do not have secular errors in energy. This often stated claim, however, applies strictly only for (quasi-)periodic systems and in the limit of small (enough) stepsize. An additional advantage of symplectic methods is that their most advanced versions are also an order of magnitude faster than any of

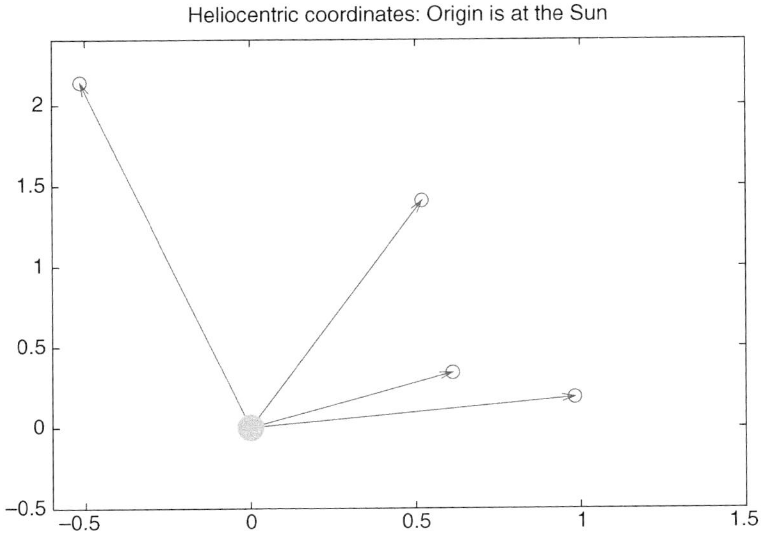

Fig. 6.1. In heliocentric coordinates every planet's motion is measured from the centre of the Sun.

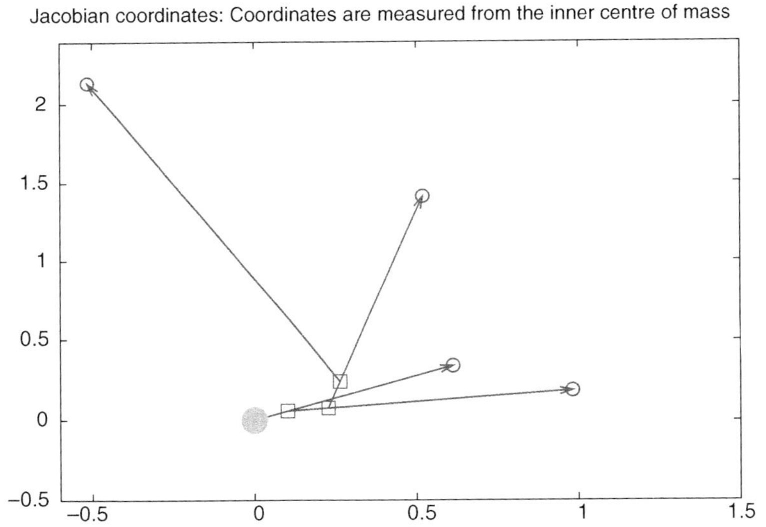

Fig. 6.2. In Jacobian coordinates every outer planet's motion is computed in a system where the radius vector's origin is at the centre of mass of the inner system.

the more classical integrators. This is the case when the problem is almost integrable, so that a major part of the solution can be treated in an analytical way.

For planetary system integrations, the basic method was outlined by Wisdom and Holman (1991), hereafter WH, and independently by Kinoshita, Yoshida and Nakai (1991), who developed a simple generalized leapfrog integration scheme. This second-order scheme works well, and has been further enhanced by the addition of a corrector (Wisdom, Holman and Touma, 1996), hereafter WHT. The symplectic corrector exploits the fact that, after many timesteps, the numerically calculated coordinates and momenta are related to the true values by a canonical transformation. One can therefore find a generating function to remove these errors.

Since and before the Wisdom–Holman scheme, there have been many other numerical schemes proposed (Yoshida, 1990; McLachlan, 1995; Chambers and Murison, 2000; Laskar and Robutel, 2001). For a review of the different numerical schemes, see Sanz-Serna (1992).

In this chapter it is first demonstrated that motion under a Hamiltonian is indeed a canonical transformation. Then it is shown that the leapfrog method is symplectic. Various generalizations of the leapfrog are introduced, the most celebrated one of which is the Wisdom–Holman method for long-time integrations of planetary motions. Subsequently, the use of correctors and the symplectic versions of many conventional numerical integration formulae are discussed. Finally, regularization by time-transformation techniques is discussed.

6.1 Basic Theory

6.1.1 Infinitesimal Canonical Transformations

Consider the canonical transformation $(p,q) \to (p_1,q_1)$ defined by the generating function

$$S = p_1 q + \epsilon \mathcal{X}(p_1,q), \tag{6.1}$$

where ϵ is an infinitesimal. Then we have

$$q_1 = \frac{\partial S}{\partial p_1} = q + \epsilon \frac{\partial \mathcal{X}(p_1,q)}{\partial p_1}, \tag{6.2}$$

$$p = \frac{\partial S}{\partial q} = p_1 + \epsilon \frac{\partial \mathcal{X}(p_1,q)}{\partial q}. \tag{6.3}$$

Solving for the new variables (p_1,q_1) gives, to first order in the infinitesimal ϵ,

$$q_1 = q + \epsilon \frac{\partial \mathcal{X}(p,q)}{\partial p}, \tag{6.4}$$

$$p_1 = p - \epsilon \frac{\partial \mathcal{X}(p,q)}{\partial q}, \tag{6.5}$$

or, since we may write $q_1 - q = dq$ and $p_1 - p = dp$,

$$\frac{q_1 - q}{\epsilon} = \frac{dq}{d\epsilon} = \frac{\partial \mathcal{X}(p,q)}{\partial p}, \tag{6.6}$$

$$\frac{p_1 - p}{\epsilon} = \frac{dp}{d\epsilon} = -\frac{\partial \mathcal{X}(p,q)}{\partial q}. \tag{6.7}$$

These are canonical equations of motion with $\mathcal{X}(p,q)$ as the Hamiltonian. If these equations are solved over a fixed interval, a finite canonical transformation emerges as a result of infinitely many infinitesimal transformations. This is actually called a *Lie transformation* (although Lie transformations are usually introduced in the form of series expansions). As the above equations are canonical (Hamiltonian), then the motion of a Hamiltonian system is equivalent to a canonical transformation. In other words: if $p(0), q(0)$ are initial values and $p(t), q(t)$ are values at the moment t, then this transformation from initial to general moment is a canonical transformation. This leads to the concept of *symplectic integration*, where all movements of the system variables are canonical transformations.

6.1.2 Lie Series

The Lie transformation was already defined above. It is, however, customary to consider these transformations in terms of Lie series.

Let $f(p,q,t)$ be an arbitrary function of canonical variables p, q and time. The Lie operator $\widehat{H}$, which gives the time derivative of f when p, q move under the Hamiltonian $H(p,q,t)$, may be defined as

$$\dot{f} = \widehat{H}f = \frac{\partial f}{\partial q}\frac{\partial H}{\partial p} - \frac{\partial f}{\partial p}\frac{\partial H}{\partial q} + \frac{\partial f}{\partial t}. \tag{6.8}$$

If the momentum of time p_0 is introduced again and the original Hamiltonian replaced by the homogeneous one,

$$H(p,q,t) \rightarrow H(p,q,q_0) + p_0,$$

then the time derivative of $f(p,q,t)$ may be written in terms of the Poisson bracket

$$\dot{f} = \{f, H\} = \sum_k \left(\frac{\partial f}{\partial q_k}\frac{\partial H}{\partial p_k} - \frac{\partial f}{\partial p_k}\frac{\partial H}{\partial q_k} \right), \tag{6.9}$$

where the q include time $t = q_0$ and the p include p_0. Further, for higher derivatives,

$$\frac{d^n f}{dt^n} = \widehat{H}^n f, \tag{6.10}$$

where the powers are defined, $\widehat{H}^n = \widehat{H}(\widehat{H}^{n-1} f)$. The Taylor expansion for f may be defined by

$$f(t_0 + h) = e^{hD} f(t_0) = e^{h\widehat{H}} f. \tag{6.11}$$

Here, in the two leftmost appearances of f, the argument simply denotes the moment of time at which the value is considered, while in the *Lie series* form,

$$e^{h\widehat{H}}f = f(p,q) + h\widehat{H}f(p,q) + \frac{h^2}{2}\widehat{H}^2 f(p,q) + \cdots, \qquad (6.12)$$

the full dependence on p and q is explicitly considered (including time as one of the q and p_0 as one of the p).

Remembering that the Poisson brackets are independent of the system of canonical variables used, one recognizes that the operator $e^{h\widehat{H}}$ is coordinate-independent. It simply moves the function forward in time under the motion defined by the Hamiltonian, so that

$$e^{h\widehat{H}}f(p(0), q(0)) = f(p(h), q(h)). \qquad (6.13)$$

6.1.3 Near-Identity Generating Function

An obvious possibility to construct canonical transformations approximating the motion is the use of the generating function, i.e. the solution of the Hamilton–Jacobi (HJ) equation.

For simplicity, consider the case of perturbed two-body motion. (The method is, however, directly useful for any system.) Once again, let

$$H(\mathbf{p}, \mathbf{r}) = \tfrac{1}{2}\mathbf{p}^2 - \frac{M}{r} + R(\mathbf{r}, t), \qquad (6.14)$$

where the notation $\mathbf{r}$ for the coordinates is used, instead of $\mathbf{q}$. The Hamilton–Jacobi equation for the solution generating function S is

$$\frac{1}{2}\left(\frac{\partial S}{\partial \mathbf{r}}\right)^2 - \frac{M}{r} + R + \frac{\partial S}{\partial t}, \qquad (6.15)$$

for which a solution is sought in the form of a power series of time,

$$S = S_0 + S_1 t + S_2 t^2 + S_3 t^3 + \cdots. \qquad (6.16)$$

Using

$$\left(\frac{\partial S}{\partial \mathbf{r}}\right)^2 = \sum_{k=0}^{\infty}\left(\sum_{\nu=0}^{k}\frac{\partial S_\nu}{\partial \mathbf{r}} \cdot \frac{\partial S_{k-\nu}}{\partial \mathbf{r}}\right) t^k$$

and

$$\frac{\partial S}{\partial t} = \sum_{k=0}^{\infty}(k+1)S_{k+1}t^k,$$

one obtains the HJ equation in the form

$$0 = \left(S_1 + \frac{1}{2}\left(\frac{\partial S_0}{\partial \mathbf{r}}\right)^2 - \frac{M}{r} + R \right) t^0$$
$$+ \sum_{k=1}^{\infty} \left((k+1) S_{k+1} + \frac{1}{2} \sum_{\nu=0}^{k} \frac{\partial S_\nu}{\partial \mathbf{r}} \cdot \frac{\partial S_{k-\nu}}{\partial \mathbf{r}} \right) t^k. \tag{6.17}$$

As the coefficients of powers of time must all vanish, recursive equations for the coefficients S_k result. First of all, there is no condition for S_0, which is thus an arbitrary function, in principle. However, to find a solution which transforms the initial coordinates and momenta $\mathbf{r}_0$, $\mathbf{p}_0$ to the new ones $\mathbf{r}(t)$, $\mathbf{p}(t)$ at (small) time t, choose

$$S_0 = \mathbf{r} \cdot \mathbf{p}_0,$$

which, in the absence of higher-order terms, gives the identity transformation. Thus

$$S_0 = \mathbf{r} \cdot \mathbf{p}_0, \tag{6.18}$$

$$S_1 = -\left(\frac{1}{2}\mathbf{p}_0^2 - \frac{M}{r} + R \right), \tag{6.19}$$

$$S_{k+1} = -\frac{1}{2(k+1)} \sum_{\nu=0}^{k} \frac{\partial S_\nu}{\partial \mathbf{r}} \cdot \frac{\partial S_{k-\nu}}{\partial \mathbf{r}}, \tag{6.20}$$

which set of equations gives the coefficients in the power series solution of the HJ equation. However, the terms very soon become extremely complicated so that one is restricted to low-order approximations. The final operation is to solve for the new coordinates and momenta from the equations

$$\mathbf{p} = \frac{\partial S}{\partial \mathbf{r}}, \quad \mathbf{r}_0 = \frac{\partial S}{\partial \mathbf{p}_0}, \tag{6.21}$$

which usually are non-linear and thus lead to implicit integration methods.

Another way to derive similar results is by Picard's iteration, which can be used for any Hamiltonian. One writes the HJ equation in the form

$$S = S_0 - \int_0^t H\left(\frac{\partial S}{\partial \mathbf{r}}, \mathbf{r} \right) \partial t, \tag{6.22}$$

where the integration is done with respect to explicit time dependence only (indicated by ∂t in this integration). Thus one starts with $S = S_0 = \mathbf{r} \cdot \mathbf{p}_0$, substitutes this into the integral, considers $\mathbf{p}_0$ and $\mathbf{r}$ as constant parameters and integrates with respect to time. The operation is then repeated, resulting in one more correct term per iteration. This method is, perhaps, the simplest if one uses computer algebra programs. If so, one should note that it may simplify the procedure if the right-hand side is expanded in powers of time adding one more term every iteration.

If the first-order approximation

$$S = S_0 + tS_1 = \mathbf{r} \cdot \mathbf{p}_0 - tH(\mathbf{p}_0, \mathbf{r}) \tag{6.23}$$

is used for the generating function, the transformation formulae

$$\mathbf{r}_0 = \frac{\partial S}{\partial \mathbf{p}_0} = \mathbf{r} - t\frac{\partial H(\mathbf{p}_0, \mathbf{r})}{\partial \mathbf{p}_0}, \quad \mathbf{p} = \frac{\partial S}{\partial \mathbf{r}} = \mathbf{p}_0 - t\frac{\partial H(\mathbf{p}_0, \mathbf{r})}{\partial \mathbf{r}} \tag{6.24}$$

result. This gives, for one integration step of stepsize h, the first-order symplectic Euler method (see below). For the case $H = \frac{1}{2}\mathbf{p}^2 - U(\mathbf{r})$ this reads

$$\mathbf{r}_1 = \mathbf{r}_0 + h\mathbf{p}_0, \tag{6.25}$$

$$\mathbf{p}_1 = \mathbf{p}_0 + h\frac{\partial U}{\partial \mathbf{r}}, \tag{6.26}$$

which is an explicit algorithm.

6.2 Composite Symplectic Methods

Composite symplectic methods can be applied if the Hamiltonian can be split into two parts,

$$H = H_0 + H_1, \tag{6.27}$$

each of which is easily integrable (solvable) if considered in isolation. What one wants to calculate is

$$\exp(h\widehat{H}) = \exp(h\widehat{H}_0 + h\widehat{H}_1), \tag{6.28}$$

which correspond to the solution of the full Hamiltonian equations of the system. As this is not generally possible, one may use instead some approximation, such as

$$\exp(h\widehat{H}) \approx \exp(h\widehat{H}_0)\exp(h\widehat{H}_1), \tag{6.29}$$

which means that one moves the system (over a timestep $= h$) first using H_0 as the Hamiltonian; then the result of that is used as initial values for the next operation, which moves the system under H_1. If the solutions for the canonical equations $\dot{p} = -\partial H_0/\partial q$, $\dot{q} = \partial H_0/\partial p$ are known in closed form, then the result of the action of the operator $\exp(h\widehat{H}_0)f(p_0, q_0)$ can be calculated: simply use the known analytical solution to advance the initial values p_0, q_0 over the timestep $= h$ to get the needed values $p(h), q(h)$. The operator $\exp(h\widehat{H}_1)$ then acts on these new values (they are considered the initial values) and the equations $\dot{p} = -\partial H_1/\partial q, \;\; \dot{q} = \partial H_1/\partial p$ are solved over the timestep h. This calculation then gives the final result of the action of the operator (6.29). After this, the steps are repeated until the required time is reached.

This simple method is known as the symplectic Euler method and it is a first-order method (in long integration). A significantly better method is obtained by using

$$\exp(h\widehat{H}) \approx \exp\left(\frac{h}{2}\widehat{H}_0\right)\exp(h\widehat{H}_1)\exp\left(\frac{h}{2}\widehat{H}_0\right), \tag{6.30}$$

which is symmetrical in time. This operator states: first move the system over a timestep of length $h/2$ using H_0 as the Hamiltonian, then advance the system over the full timestep h under the Hamiltonian H_1 and follow this again by a half-timestep using H_0. This operator can be obtained from (6.29) by concatenating it with one in which the order of application of H_0 and H_1 is reversed (and writing $h/2$ for h). It is possible to show (Yoshida, 1990) that

$$\exp\left(\frac{h}{2}\widehat{H}_0\right)\exp(h\widehat{H}_1)\exp\left(\frac{h}{2}\widehat{H}_0\right) = \exp(h(\widehat{H} + \widehat{H}_{err})), \tag{6.31}$$

where

$$H_{err} = \frac{h^2}{12}\left\{\{H_0, H_1\}, H_1 + \frac{1}{2}H_0\right\} + O(h^4) \tag{6.32}$$

is an error Hamiltonian. Thus the approximation (generalized leapfrog) has an error of $O(h^3)$ in one step and $O(h^2)$ when repeated to advance over a given interval of time. In addition, if H_1 is small, say $H_1 = O(\epsilon)$, then the global error becomes $= O(\epsilon h^2)$.

As we see from these simple developments, the symplectic integration resembles the composite methods (8.27). In fact it is true that every composite method is symplectic when applied to Hamiltonian equations.

6.2.1 Leapfrog

Once more, let the Hamiltonian be given by

$$H = \frac{\mathbf{p}^2}{2} - U(\mathbf{r}), \tag{6.33}$$

and let us use the operator

$$\exp(\tfrac{1}{2}h\widehat{T})\exp(h(-\widehat{U}))\exp(\tfrac{1}{2}h\widehat{T}). \tag{6.34}$$

Here $T = \frac{1}{2}\mathbf{p}^2$ and the operator $\exp(\frac{1}{2}h\widehat{T})$ simply implies linear free-particle motion and transforms only the coordinates $\mathbf{r}_{1/2} = \mathbf{r}_0 + \frac{1}{2}hp_0$. The second operator $\exp(h(-\widehat{U}))$ is also simple to calculate because the result is a velocity jump $\mathbf{p}_1 = \mathbf{p}_0 + h\partial U/\partial\mathbf{r}$. When these two operations are followed by the final linear motion $\exp(\frac{1}{2}h\widehat{T})$, the leapfrog algorithm ensues:

$$\mathbf{r}_{1/2} = \mathbf{r}_0 + \tfrac{1}{2}h\mathbf{p}_0, \tag{6.35}$$

$$\mathbf{p}_1 = \mathbf{p}_0 + h\frac{\partial U}{\partial \mathbf{r}}, \tag{6.36}$$

$$\mathbf{r}_1 = \mathbf{r}_{1/2} + \tfrac{1}{2}h\mathbf{p}_1. \tag{6.37}$$

As for all symplectic methods, this can be shown to have a surrogate Hamiltonian

$$H_{lf} = H + h^2H_2 + h^4H_4 + \cdots = H + H_{err}, \tag{6.38}$$

for which the leapfrog (formally) gives correct results.

The existence of the error Hamiltonian is the reason why symplectic methods do not have secular errors. Strictly speaking, this property requires that the solutions are quasi-periodic, so that the value of H_{err} varies in a periodic fashion. In systems which significantly change their structure during the evolution, one may see corresponding variations in energy error.

However, the series is usually only semi-convergent, and in critical interactions the terms H_2, H_4, etc. may be large and thus the computed orbit may lose all resemblance to the true solution of the problem.

The error Hamiltonian always depends on the stepsize and this fact imposes an important restriction on the use of symplectic integrators in general: *the stepsize must be kept constant*. Otherwise one would be changing the Hamiltonian and the result would again be a secular development of error.

6.2.2 Higher-Order Leapfrogs

It is possible to construct methods in which the error is of higher order by combining leapfrog steps of different size. However, it has been shown that this can only be achieved if some steps are negative (i.e. go backwards). A well-known example is the fourth-order method, which can be written as a composite of three leapfrog steps, one of which is negative. Let

$$\mathbf{L}(h) = \mathbf{X}(h/2)\mathbf{V}(h)\mathbf{X}(h/2) \tag{6.39}$$

be a leapfrog, where the notation **X** for the coordinate motion operator and **V** for the velocity jump have been used, and the argument is the stepsize. The fourth-order method can be given by

$$\mathbf{L}_4(h) = \mathbf{L}(w_1 h)\mathbf{L}(w_0 h)\mathbf{L}(w_1 h), \tag{6.40}$$

which is a time-symmetrical product of leapfrog operators. The method is of fourth order if $w_1 : w_0 = 1 : -2^{1/3}$ or, due to the normalization $w_0 = 1 - 2w_1$,

$$w_1 = 1/(2 - 2^{-1/3}), \quad w_0 = -2^{-1/3}/(2 - 2^{-1/3}). \tag{6.41}$$

By writing the operator in full product form there results

$$\mathbf{L}_4(h) = \mathbf{X}(x_1 h)\mathbf{V}(w_1 h)\mathbf{X}(x_0 h)\mathbf{V}(w_0 h)\mathbf{X}(x_0)\mathbf{V}(w_1 h)\mathbf{X}(x_1 h), \tag{6.42}$$

where $x_1 = w_1/2$ and $x_0 = (w_0 + w_1)/2$.

As shown by Yoshida (1990), it is possible to construct higher-order methods of the general form

$$L_{h.o.}(h) = \mathbf{L}(w_m h)\mathbf{L}(w_{m-1} h) \cdots \mathbf{L}(w_0 h) \cdots \mathbf{L}(w_m h), \tag{6.43}$$

which can also be written

$$\mathbf{L}_{h.o.}(h) = \mathbf{X}(x_m h)\mathbf{V}(w_m h) \cdots \mathbf{X}(x_0 h)\mathbf{V}(w_0)\mathbf{X}(x_0 h) \cdots \mathbf{V}(w_m h)\mathbf{X}(x_m h), \tag{6.44}$$

where $x_m = w_m/2,\ x_{m-1} = (w_m + w_{m-1})/2, \ldots, x_0 = (w_1 + w_0)/2$.

To obtain a sixth-order formula, one needs $m = 3$, and for eighth order, it is necessary to take $m = 7$. Since the number of force evaluations per step is $2m + 1$, it seems that, in practice, it is questionable if formulae of order higher than sixth are useful. At order six there are several possible solutions, but the one with smallest error is

$$\begin{aligned} w_1 &= -1.17767998417887, \\ w_2 &= 0.235573213359357, \\ w_3 &= 0.784513610477560, \\ w_0 &= 1 - 2(w_1 + w_2 + w_3). \end{aligned} \tag{6.45}$$

These higher-order formulae are not restricted to the simple leapfrog representable by the x-motion and v-jump operators, but they are valid also for the 'generalized leapfrogs' discussed below.

A FORTRAN Routine for the Fourth-Order Method

```
      subroutine L4(h,time)
      implicit real*8 (a-h,M,o-z)
      real*8 hc(4),hd(3)
c      These should be given in a table
```

```
      s3=2.d0**(1.d0/3.d0)
      hx(1)=h*.5d0/(2-s3)
      hx(4)=hx(1)
      hx(2)=hx(1)*(1-s3)
      hx(3)=hx(2)
      hw(1)=2*hx(1)
      hw(3)=hw(1)
      hw(2)=-s3*h/(2-s3)

c     Computation  (x's and v's are supposed to be in a common block)
      do k=1,3
     call Xmotion(hx(k)) ! linear motion x=x+hc(k)*v
     call Vmotion(hw(k)) ! velocity jump v=v+hd(k)*F
      end do
     call Xmotion(hx(4)) ! final linear motion to the end of step
     hcs=hx(1)+hx(2)+hx(3)+hx(4) ! =\sum_k
     hd(k)
     time=time+hcs
     return
     end
```

An Order-Six Leapfrog Code (Method of Yoshida)

```
      subroutine Y6A(h)
      implicit real*8 (a-h,M,o-z)
      real*8 hc(8),hd(7),c(8),d(7)
      save

      c( 1 )=  0.3922568052387800d0
      d( 1 )=  0.7845136104775600d0
      c( 2 )=  0.5100434119184585d0
      d( 2 )=  0.2355732133593570d0
      c( 3 )= -0.4710533854097566d0
      d( 3 )=  -1.177679984178870d0
      c( 4 )=  6.875316825251809 3d-02
      d( 4 )=   1.315186320683906d0
      c( 5 )=  6.8753168252518093d-02
      d( 5 )=  -1.177679984178870d0
      c( 6 )= -0.4710533854097566d0
      d( 6 )=  0.2355732133593570d0
      c( 7 )=  0.5100434119184585d0
      d( 7 )=  0.7845136104775600d0
      c( 8 )=  0.3922568052387800d0
        do i=1,7
```

```
    hc(i)=c(i)*h
    hd(i)=d(i)*h
    end do
    hc(8)=c(8)*h
   do k=1,7
  call Xmotion(hc(k))
  call Vjump(hd(k))
   end do
  call Xmotion(hc(8))
  return
  end
```

6.2.3 Generalized Leapfrog

When the Hamiltonian is

$$H = H_0 + H_1, \tag{6.46}$$

where H_0 and H_1 are both individually integrable, then the *generalized leapfrog* can be written

$$e^{(1/2)h\widehat{H}_0} e^{h\widehat{H}_1} e^{(1/2)h\widehat{H}_0}. \tag{6.47}$$

If we have, for example, $H_0 = \mathbf{p}^2/2 - M/r + p_0$ as a Keplerian Hamiltonian and $H_1 = R(\mathbf{r}, t)$ as a small perturbation, then the operator $\exp(\frac{1}{2}h\widehat{H}_1)$ produces the so-called *momentum jump*

$$\delta\mathbf{p} = \frac{h}{2}\frac{\partial R}{\partial \mathbf{r}}, \quad \delta p_0 = \frac{h}{2}\frac{\partial R}{\partial t}, \tag{6.48}$$

and the operator $\exp(h\widehat{H}_0)$ produces pure Kepler motion (plus advancement of time) over a timestep of length h. Note that the momentum jump is most often written in terms of velocities and is then called the *velocity jump*. This formulation is often used for almost-integrable systems, i.e. the H_1 part is a small perturbation and most of the system is treated exactly by the 'unperturbed' part H_0. An example of an orbit computed with this generalized leapfrog is illustrated in Figure 6.3. This is an interesting asteroid orbit that is a so-called quasi-satellite trajectory around Venus (Mikkola et al., 2004).

One may also observe that the quantity p_0 is not actually needed; however, it can still be useful in checking the accuracy of the method in those (rather common) cases where the system

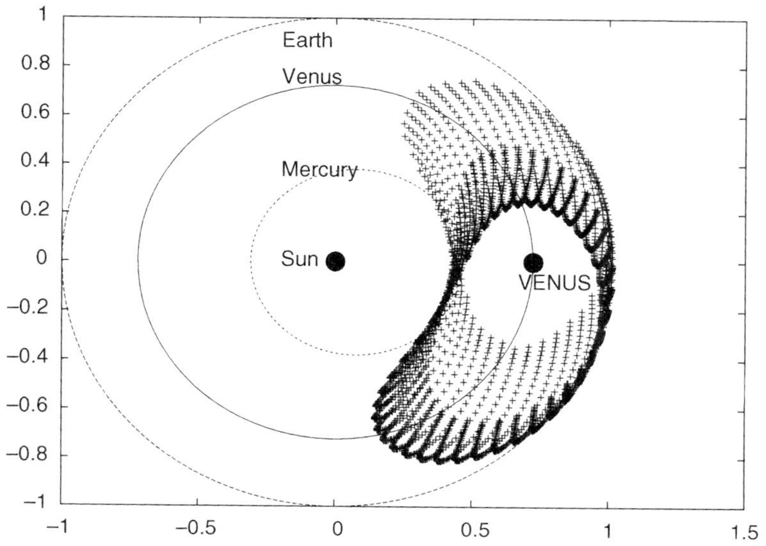

Fig. 6.3. Motion of the asteroid 2002 VE68 in the coordinate system that rotates with Venus. This was computed with a symplectic integrator and plotted for 150 years. According to the computation this asteroid has been near Venus for about 7000 years and will stay there still for about 500 years.

does not have any integrals of motion. The value of $H_0 + H_1 + p_0$, however, is an integral which can be used as an accuracy check of the numerical method, although it does not give any dynamical information.

6.2.4 Delta-Function Formalism

Instead of Lie operators one may use the δ-function formulation of Wisdom and Holman. Here the original Hamiltonian $H = H_0 + H_1$ is replaced by

$$H_\delta = H_0 + h\delta_h(t)H_1, \tag{6.49}$$

where $\delta_h(t)$ is a periodic delta-function,

$$\delta_h(t) = \sum_{k=1}^{\infty} \delta(t + h(k - \tfrac{1}{2})).$$

This function is sometimes called the δ-function fence because it acts at regular times separated by the timestep h.

Let us start the integration at time $t = 0$. Up to time $t = \frac{1}{2}h$, the $\delta_h(t)$-function is zero and the motion is defined by the Hamiltonian H_0. When the δ-function acts at $t = \frac{1}{2}h$, everything else but the H_1 part is zero in comparison. Consequently, in an infinitesimal interval of time dt, the Hamiltonian is $= (h/dt)H_1$ and, assuming $H_1 = R(\mathbf{r}, t)$, this results in the momentum jumps

$$\Delta \mathbf{p} = -dt \frac{h}{dt}\frac{\partial H_1}{\partial \mathbf{r}} = -h\frac{\partial R(\mathbf{r},t)}{\partial \mathbf{r}},$$

because the time interval is infinitesimally small and the coordinates thus remain constant. Similarly, if the momentum of time p_0 is included in the formalism, $\Delta p_0 = -h\partial H_1 \partial t$.

In some cases, the perturbation part H_1 is such that part of it is small and slowly varying but expensive to evaluate. Let us write this as $H = R_1(\mathbf{r},t) + R_2(\mathbf{r},t)$. Then one may adopt a two-fence formulation by taking

$$H_\delta = H_0 + h_1\delta_{h_1}(t - \tfrac{1}{2}h_1)R_1(\mathbf{r},t) + h_2\delta_{h_2}(t - \tfrac{1}{2}h_2)R_2(\mathbf{r},t), \quad (6.50)$$

where the steplengths of the δ-functions may be chosen in a suitable way depending on the behaviour of the two perturbations. Naturally this method is not restricted to just two different terms but one may utilize several of them. Again, the system moves according to the unperturbed Hamiltonian H_0, except when one or other of the δ-functions act, causing a velocity jump.

It may be of some interest to note that the basic formulation (6.49) can be replaced by

$$H = h\delta_h(t)H_0 + H_1. \quad (6.51)$$

When the δ-function acts, the result is an instantaneous motion under the Hamiltonian H_0. The final algorithm is just the same as that following from (6.49). One could even have the δ-function fences in front of both parts of the Hamiltonian, i.e. $H = h\delta_h(t)H_0 + h\delta_h(t - \frac{1}{2}h)H_1$, without changing the result.

6.2.5 Lyapunov Exponents

An important special case of the generalized leapfrog is the Wisdom–Holman method (Wisdom and Holman, 1991), for which the basic theory was already given in Section 6.1. This consists of splitting the Hamiltonian of the Solar System (or of any similar hierarchical system) into a sum of two-body Hamiltonians and a perturbing function which depends only on the coordinates. This can be done in Jacobian coordinates and the result for an n-planet system has the form

$$H = T_{cm} + \sum_{j=1}^{n} K_j + R(\mathbf{r}_1, \ldots, \mathbf{r}_n), \quad (6.52)$$

where T_{cm} is the kinetic energy of the centre of mass, the K_j are the Kepler-motion Hamiltonians for the planets and R is the perturbing function. In Jacobian coordinates each planet's position is taken relative to the centre of mass of the 'lower' planets plus the Sun. In these variables, the two-body orbits are typically better approximations to the real motion than in heliocentric coordinates. The transformation is

$$\mathbf{r}_i^J = \mathbf{r}_i - \mathbf{X}_i, \qquad \mathbf{X}_i = \frac{\sum_{k=0}^{i-1} m_k \mathbf{r}_k}{\sum_{k=0}^{i-1} m_k}, \tag{6.53}$$

and the Hamiltonian is

$$H_J = \frac{\mathbf{p}_{cm}^2}{2M_{cm}} + \sum_{i=0}^{n} \frac{\tilde{\mathbf{p}}_i^2}{2\mu_i} - \sum_{i<j} \frac{m_i m_j}{r_{ij}}, \tag{6.54}$$

where μ_i is the reduced mass,

$$\mu_i = \frac{m_i M_{i-1}}{M_i}, \quad M_j = \sum_{k=0}^{j} m_k. \tag{6.55}$$

The centre-of-mass-related quantities $\mathbf{p}_{cm}$ and M_{cm} are

$$\mathbf{p}_{cm} = \sum \mathbf{p}_i, \quad M_{cm} = M_n. \tag{6.56}$$

The relative distances r_{ij} are expressed in terms of the Jacobian coordinates $\mathbf{r}_i^J$ and the Jacobian momenta are

$$\tilde{\mathbf{p}}_i = \mu_i \dot{\mathbf{r}}^J. \tag{6.57}$$

The two-body Hamiltonians K_j are

$$K_j = \frac{\tilde{\mathbf{p}}_j^2}{2\mu_j} - \frac{m_j M_{j-1}}{r_j^J} \tag{6.58}$$

and the perturbing function is

$$R = \sum_j \frac{m_j M_{j-1}}{r_j^J} - \sum_{i<j} \frac{m_i m_j}{r_{ij}}. \tag{6.59}$$

The algorithm proceeds by first moving the planets along Keplerian ellipses over a half-timestep applying

$$\exp\left(\frac{1}{2} h \sum_{j=1}^{n} \mathbf{K}_j\right),$$

followed by

$$\exp(h\mathbf{R}),$$

corresponding to the momentum increments

$$\delta \mathbf{p}_j = -h \frac{\partial R}{\partial \mathbf{r}_j}, \tag{6.60}$$

which are to be added to the momenta. Normally these momentum increments are carried out in the form of velocity jumps. The simple rule is that one writes the equations of motion for each planet in the form

$$\ddot{\mathbf{r}}_j = -M_j \mathbf{r}_j / r_j^3 + \mathbf{F}_j, \qquad (6.61)$$

where the $\mathbf{F}_j$ are the perturbing accelerations

$$\mathbf{F}_j = -\frac{1}{m_j} \frac{\partial R}{\partial \mathbf{r}_j^J}, \qquad (6.62)$$

which increment the velocities by the amounts

$$\delta \mathbf{v}_j = h \mathbf{F}_j. \qquad (6.63)$$

For purposes of programming of these formulae one should note that the velocities, as well as accelerations, satisfy the same transformation formulae (6.53) as the coordinates. The details of the algorithm, with collections of formulae, are given below in the subsection on 'Keplerian motion with variations'.

Tangent Map in Symplectic Integration

The Lyapunov exponent λ is a measure of the degree of sensitivity of the system to changes of initial values and/or external perturbation. It may be defined as

$$\lambda = \lim_{t \to \infty} \frac{\ln(|\delta \mathbf{r}|)}{t}, \qquad (6.64)$$

where $\delta \mathbf{r}$ is a difference of two orbits starting very near each other. Since a constant factor in $\delta \mathbf{r}$ is irrelevant in this definition, the differentials of the position vector(s) can also be used for it. For simplicity, consider (first) the details in the case of a simple perturbed two-body motion.

The system under consideration has a Hamiltonian of the form

$$H = K + R, \qquad (6.65)$$

where

$$K = \tfrac{1}{2} \mathbf{v}^2 - M / |\mathbf{r}| \qquad (6.66)$$

is a Keplerian Hamiltonian (in this notation the momentum and velocity are equal) and

$$R = R(\mathbf{r}, t) \qquad (6.67)$$

is a perturbing function.

The equations of motion may thus be written

$$\dot{\mathbf{r}} = \mathbf{v}, \tag{6.68}$$

$$\dot{\mathbf{v}} = -\frac{M\mathbf{r}}{r^3} - \frac{\partial R}{\partial \mathbf{r}}. \tag{6.69}$$

The equations of motion for the difference between two adjacent motions can be obtained by differentiation. The resulting equations are known as the *variational equations*:

$$\delta\dot{\mathbf{r}} = \delta\mathbf{v}, \tag{6.70}$$

$$\delta\dot{\mathbf{v}} = -M\left[\frac{\delta\mathbf{r}}{r^3} - \frac{3\mathbf{r}}{r^5}(\mathbf{r}\cdot\delta\mathbf{r})\right] - \frac{\partial^2 R}{\partial \mathbf{r}^2}\cdot\delta\mathbf{r}. \tag{6.71}$$

For the numerical solution of the above equations (from (6.68) to (6.71)), the generalized leapfrog (WH) algorithm will be considered here. It is more convenient to compute the variations $\delta\mathbf{r}$ and $\delta\mathbf{v}$ by *differentiation of the leapfrog algorithm*, instead of considering the solution of the differential equations (6.71). The resulting map is known as the tangent map of the motion.

To obtain expressions for the advancement of the motion variations, it is first necessary to write the collection of equations in the order required by the algorithm. The sequential differentiation of the latter equations produces the set of equations (the algorithm) to advance the motion variations. In the next subsections, first, the relevant equations for advancing the coordinates and velocities are listed. Then, these equations are differentiated to obtain expressions for the advancement of the tangent map. This procedure is equivalent to solving the variational equations and providing the variations $\delta\mathbf{r}$ and $\delta\mathbf{v}$ as functions of time.

Keplerian Motion with Variations

Here we consider the equations for the advancement of Keplerian motion and its variations over a timestep of length h. The particular formulation chosen is completely 'universal' in the sense that it is valid for all types of two-body orbits including even those with repulsive forces (by change of sign of the mass). The formulae for the motion itself are listed in equations (6.72). Starting with the initial position $\mathbf{r}_0$ and velocity $\mathbf{v}_0$, one first computes the quantities β, η_0 and ζ_0. The equation $t(X) = h$ is then solved for X, where $X = \int dt/r$ satisfies the 'universal' form of Kepler's equation ('Hauptgleichung' of Stumpff (1959)). The solution of the Hauptgleichung can be performed efficiently by extrapolating the solution from previous steps and using some high-order iteration formula to provide a specified precision. Knowing the value of X, one can evaluate the c- and G-functions as well as f and g and their time derivatives. The new positions $\mathbf{r}$ and velocities $\mathbf{v}$ are obtained from equations (6.72).

Kepler motion

$$\left\{\begin{aligned}
\beta &= 2M/r_0 - \mathbf{v}_0^2 \ \ (= M/a),\\
\eta_0 &= \mathbf{r}_0 \cdot \mathbf{v}_0,\\
\zeta_0 &= M - \beta r_0,\\
t(X) &= r_0 X + \eta_0 G_2 + \zeta_0 G_3 = h \quad \text{(solve for } X\text{)},\\
r &= r_0 + \eta_0 G_1 + \zeta_0 G_2,\\
f &= 1 - M G_2 / r_0,\\
g &= t - M G_3,\\
\dot{f} &= -M G_1/(r_0 r),\\
\dot{g} &= 1 - M G_2 / r,\\
\mathbf{r} &= f \mathbf{r}_0 + g \mathbf{v}_0,\\
\mathbf{v} &= \dot{f} \mathbf{r}_0 + \dot{g} \mathbf{v}_0.
\end{aligned}\right. \tag{6.72}$$

Variations

$$\left\{\begin{aligned}
\delta r_0 &= \mathbf{r}_0 \cdot \delta \mathbf{r}_0 / r_0,\\
\delta \beta &= -2M \delta r_0 / r_0^2 - 2\mathbf{v}_0 \cdot \delta \mathbf{v}_0,\\
\delta \eta_0 &= \mathbf{v}_0 \cdot \delta \mathbf{x}_0 + \mathbf{r}_0 \cdot \delta \mathbf{v}_0,\\
\delta \zeta_0 &= -\beta \delta r_0 - r_0 \delta \beta,\\
G_{3\,\beta} &= (3G_5 - X G_4)/2,\\
G_{2\,\beta} &= (2G_4 - X G_3)/2,\\
G_{1\,\beta} &= (G_3 - X G_2)/2,\\
t_\beta &= \eta_0 G_{2\,\beta} + \zeta_0 G_{3\,\beta},\\
\delta X &= -(X \delta r_0 + G_2 \delta \eta_0 + G_3 \delta \zeta_0 + t_\beta \delta \beta)/r,\\
\delta G_1 &= G_0 \delta X + G_{1\,\beta} \delta \beta,\\
\delta G_2 &= G_1 \delta X + G_{2\,\beta} \delta \beta,\\
\delta G_3 &= G_2 \delta X + G_{3\,\beta} \delta \beta,\\
\delta r &= \delta r_0 + G_1 \delta \eta_0 + G_2 \delta \zeta_0 + \eta_0 \delta G_1 + \zeta_0 \delta G_2,\\
\delta f &= M G_2 \delta r_0 / r_0^2 - M \delta G_2 / r_0,\\
\delta g &= -M \delta G_3,\\
\delta \dot{f} &= -M \delta G_1 / (r_0\, r) + M G_1 (\delta r_0 / r_0 + \delta r / r)/(r r_0),\\
\delta \dot{g} &= -M \delta G_2 / r + M G_2 \delta r / r^2,\\
\delta \mathbf{r} &= f \delta \mathbf{r}_0 + g \delta \mathbf{v}_0 + \mathbf{r}_0 \delta f + \mathbf{v}_0 \delta g,\\
\delta \mathbf{v} &= \dot{f} \delta \mathbf{r}_0 + \dot{g} \delta \mathbf{v}_0 + \mathbf{r}_0 \delta \dot{f} + \mathbf{v}_0 \delta \dot{g}.
\end{aligned}\right. \tag{6.73}$$

If the values of the variations of position and velocity are $\delta \mathbf{r}_0$ and $\delta \mathbf{v}_0$ at the beginning of a Keplerian timestep, then their values at the end of the step can be obtained by differentiating the collection of formulae for the advancement of the Kepler motion. All the necessary expressions are collected above in equations (6.73).

In that system, $G_n = G_n(\beta, X)$; the subscript β indicates partial differentiation with respect to that quantity. For the solution of the equation

$$t(X) = h,$$

the following derivatives are useful:

$$t' = r, \quad t'' = \eta = \eta_0 G_0 + \zeta_0 G_1.$$

Here Halley's method,

$$\delta t = t(X) - h, \quad X_{new} = X - \delta t/[t' - \delta t\, t''/(2t')],$$

for improving the estimate of X has been adopted, initially obtained by extrapolation from previous steps.

Velocity Increments with Variations

The velocity increments are simply

$$\Delta \mathbf{v} = h\mathbf{F} = -h\frac{\partial R}{\partial \mathbf{r}}, \tag{6.74}$$

where $\mathbf{F}$ is the perturbing acceleration. Next, the increments for the variations are obtained from

$$\delta \Delta \mathbf{v} = h\delta \mathbf{F} = h\frac{\partial \mathbf{F}}{\partial \mathbf{r}} \cdot \delta \mathbf{r} = -h\frac{\partial^2 R}{\partial \mathbf{r}^2} \cdot \delta \mathbf{r}. \tag{6.75}$$

At the moment of the velocity increments, one thus adds $\Delta \mathbf{v}$ to the velocity, and, similarly, $\delta \Delta \mathbf{v}$ is to be added to the tangent-map vector $\delta \mathbf{v}$.

As an example, consider the motion of an asteroid under the perturbation of Jupiter. The equation of motion is

$$\ddot{\mathbf{r}} = -M_\odot \mathbf{r}/r^3 + m_J(\mathbf{d}/d^3 - \mathbf{r}_J/r_J^3), \tag{6.76}$$

where $\mathbf{d} = \mathbf{r}_J - \mathbf{r}$ and thus $\mathbf{F} = m_J(\mathbf{d}/d^3 - \mathbf{r}_J/r_J^3)$. The variation of the perturbing acceleration takes the form

$$\delta \mathbf{F} = -m_J \left[\frac{\delta \mathbf{r}}{d^3} - 3\frac{\mathbf{d}(\mathbf{d} \cdot \delta \mathbf{r})}{d^5} \right]. \tag{6.77}$$

In Solar System computations, one usually uses the Jacobian coordinates, which may be computed by the simple linear (and recursive) transformation (subscript 0 referring to the Sun)

$$\mathbf{r}_i^J = \mathbf{r}_i - \mathbf{X}_{i-1},$$

$$M_n = \sum_{i=0}^{n} m_n,$$

$$\mathbf{X}_n = \frac{\sum_{i=0}^{n} m_k \mathbf{r}_k}{\sum_{k=0}^{i} m_k} = (\mathbf{X}_{n-1} M_{n-1} + \mathbf{r}_k m_n)/M_n. \tag{6.78}$$

Note that this system can be inverted immediately, for the recursion works in both directions. Furthermore, as this transformation is linear, it applies to all the derivatives of the coordinates. Often, it is simpler to program the computation of forces in heliocentric rather than Jacobian coordinates. One may thus transform from Jacobian to heliocentric coordinates, evaluate the forces (and/or their variations) and transform the results back to Jacobian coordinates. Therefore, when discussing the evaluation of the variation, one needs only to show explicitly the operations in heliocentric coordinates. The perturbing acceleration (in heliocentric coordinates) is

$$\mathbf{F}_i = \sum_{j \neq 0, i} m_i (\mathbf{r}_{ij}/r_{ij}^3 - \mathbf{r}_{0j}/r_{0j}^3), \tag{6.79}$$

where $\mathbf{r}_{ij} = \mathbf{r}_j - \mathbf{r}_i$. This is a linear combination of terms of the form $\mathbf{r}_{ij}/r_{ij}^3$. The variations of such terms are analogous to the one in (6.77) so that

$$\delta \mathbf{r}_{ij} = \delta \mathbf{r}_j - \delta \mathbf{r}_i,$$

$$\delta \mathbf{F}_i = \sum_{j \neq 0, i} m_j \left[\left(\frac{\delta \mathbf{r}_{ij}}{r_{ij}^3} - \frac{3\mathbf{r}_{ij}(\mathbf{r}_{ij} \cdot \delta \mathbf{r}_{ij})}{r_{ij}^5} \right) - \left(\frac{\delta \mathbf{r}_{0j}}{r_{0j}^3} - \frac{3\mathbf{r}_{0j}(\mathbf{r}_{0j} \cdot \delta \mathbf{r}_{0j})}{r_{0j}^5} \right) \right]. \tag{6.80}$$

Following the evaluation of the heliocentric variations, one transforms to Jacobian coordinates using the formulae given above (written for coordinates). These Jacobian coordinate contributions are then added to the velocities (in Jacobian coordinates) and the velocity increment is completed. Corresponding expressions for other cases are usually simple to obtain and further details are omitted here.

Computing Orbital Motion and Variations

The directions for using these formulae with the WH algorithm are as follows.

First, transform the initial values (positions and velocities) to Jacobian coordinates by means of (6.53) and adopt arbitrary (but not all zero) starting values for the variations $\delta\mathbf{r}$ and $\delta\mathbf{v}$. Then, continue according to:

1. Use (6.72) to advance $\mathbf{r}$ and $\mathbf{v}$ over the Kepler timestep of length $h/2$. To obtain the c- and G-functions, use (2.137); to advance the variations, the formulae (6.73) are to be applied.
2. The values of the expressions (6.79) are added to the velocities after multiplication by the timestep h.
3. Similarly, the product of h and the expressions (6.80) are to be added to the variations $\delta\mathbf{v}$ of the velocities.
4. Next, the system (6.72) is used to advance $\mathbf{r}$ and $\mathbf{v}$ over the full Kepler timestep h. This is followed by the above incremental steps 2 and 3 and the algorithm continues until the next output event.
5. Before output, a half-timestep Kepler motion, according to step 1, is to be carried out.

From the results of the computation, one concurrently obtains the orbits of the bodies as well as the associated Lyapunov exponents:

$$\lambda_i = \lim_{t\to\infty} \ln(|d\mathbf{r}_i|)/t, \tag{6.81}$$

while the e-folding time τ_e for the motion is simply

$$\tau_{e\,i} = 1/\lambda_i. \tag{6.82}$$

Next, we turn to the computation of the state transition matrix, i.e. the matrix $\mathbf{S}(t)$, which gives the variations at the moment t in terms of the initial values,

$$\delta\mathbf{q}(t) = \mathbf{S}(t)\,\delta\mathbf{q}(0), \tag{6.83}$$

where, in obvious notation, $\mathbf{q} = (\delta x, \delta y, \delta z, \delta v_x, \delta v_y, \delta v_z)^t$.

This matrix is useful, e.g. in orbit determination from observations. It can be obtained by computing six linearly independent solutions for the variational equations: one adopts for initial values of $\mathbf{q}$ the unit vectors

$$\mathbf{q}(0) = \mathbf{e}_k \quad (k = 1, 2, \ldots, 6) \tag{6.84}$$

where the kth component $\mathbf{e}_k$ is one, while the other components are zeros. The corresponding solution vectors, obtained by the above algorithm, are then columns of the matrix $\mathbf{S}$, which is initially $\mathbf{S}(0) = \mathbf{I}$, a unit matrix.

For purposes of checking a code written along the lines described here, one can use the following. The partial derivatives of the solution of (6.68) with respect to initial values form a basic set of solutions for these variational equations. Thus the differentials

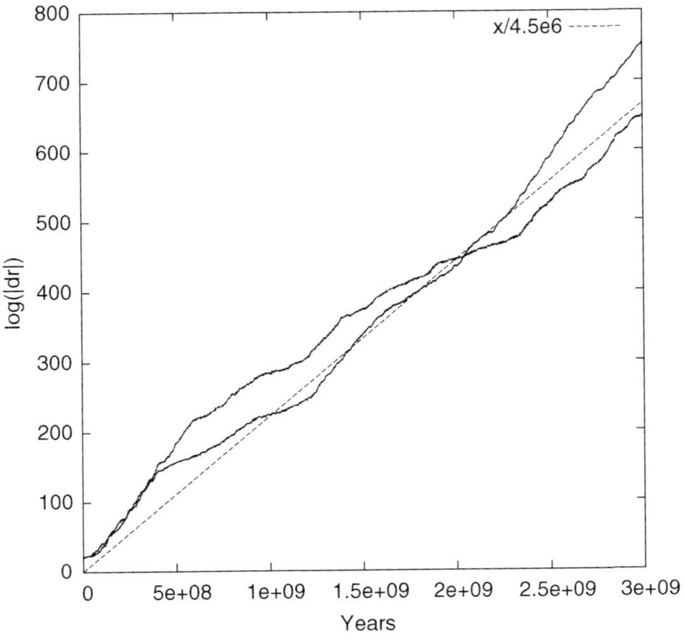

Fig. 6.4. Evolution of $\log(|d\mathbf{r}|)$ in the Solar System with two different symplectic integrators. The straight dashed line represents the behaviour when the e-folding time is 4.5 Myr.

of any constant of motion of (6.68) are constants of motion of the variational equations (6.71). If the Hamiltonian is conservative, the variation

$$\delta H = (\partial H/\partial \mathbf{r}) \cdot \delta \mathbf{r} + (\partial H/\partial \mathbf{v}) \cdot \delta \mathbf{v} = \text{constant}, \tag{6.85}$$

a result which can be used as a check in the computation. Similarly the variations of the angular momentum,

$$\delta L_{ij} = \delta x_i \dot{x}_j + x_i \delta \dot{x}_j - \delta \dot{x}_i x_j - \dot{x}_i \delta x_j, \tag{6.86}$$

should remain constant if the angular momentum itself is constant. However, one should note that, usually, the variations grow rapidly in numerical value. Consequently, after a long integration, small relative errors in the variations may be sufficient to destroy the (expected) conservation of these quantities.

In Figure 6.4 there are results for the evolution of the size of the solutions of the variational equations. The two alternatives shown are the results with the basic WH method and with the alternative stepsize method with the remainder term given in equation (6.162) in Section 6.5.5. Here the e-folding time seems to be $\approx 4.5 \times 10^6$ years. Often this number has been obtained to be about 6 Myr. The difference is likely because these results were obtained by including also the planet Mercury, which has normally been incorporated into the Sun. The reason is Mercury's short orbital period.

6.2.6 Individual Accuracy Checks

Symplectic codes of the WH type usually lack a reliable accuracy test for the integration of orbits of asteroids, as these bodies have no influence on the total energy of the system. Mikkola and Innanen (2002) have suggested a simple device which helps in this respect.

To construct individual accuracy checks it is convenient to consider the equations of motion in the inertial (centre-of-mass) coordinate system. Here it is possible to use the individual Hamiltonians

$$H_i = \frac{\mathbf{v}_i^2}{2} - \sum_{j \neq i} \frac{m_j}{|\mathbf{r}_i - \mathbf{r}_j|}, \tag{6.87}$$

where the velocities $\mathbf{v}_i$ can be used as momenta, and the coordinates of all the other bodies $\mathbf{r}_j$ are considered to be functions of time only. As is easy to see, the equations of motion so obtained are the correct ones.

A simple way to construct the individual accuracy checks is now available. The time derivative of the individual Hamiltonians H_i can be written

$$\frac{dH_i}{dt} = \frac{\partial H_i}{\partial t} = \sum_{j \neq i} \frac{\partial H_i}{\partial \mathbf{r}_j} \cdot \mathbf{v}_j = -\sum_{j \neq i} \frac{\mathbf{r}_i - \mathbf{r}_j}{|\mathbf{r}_i - \mathbf{r}_j|^3} \cdot \mathbf{v}_j. \tag{6.88}$$

This equation can be integrated along with the velocities. The velocity jump is actually the solution to the differential equation of the form

$$\delta \dot{\mathbf{v}} = \mathbf{F}, \tag{6.89}$$

with zero initial conditions and constant $\mathbf{F}$, i.e.

$$\delta \mathbf{v} = t\mathbf{F}, \tag{6.90}$$

over the time interval of length h.

One sees that the changes of the individual energies H_i are

$$\delta H_i = -h \sum_{j \neq i} \frac{\mathbf{r}_i - \mathbf{r}_j}{|\mathbf{r}_i - \mathbf{r}_j|^3} \cdot \langle \mathbf{v}_j \rangle, \tag{6.91}$$

where the $\langle \mathbf{v}_j \rangle$ are the mean velocities over the velocity jump in question. As one is working with the Jacobian coordinates and corresponding velocities, these can be conveniently evaluated by first computing the mean Jacobian velocities

$$\left\langle \mathbf{v}_j^J \right\rangle = \mathbf{v}_j^J(0) + \frac{h}{2}\mathbf{F}_j \tag{6.92}$$

($\mathbf{v}_j^J(0)$ = velocity before adding the jump), which are then transformed to the centre-of-mass system to obtain those needed in (6.91).

Finally, computation of the values of H_i according to (6.87) allows comparison with the integrated value. This check is especially useful for massless bodies, but can also be used for massive ones.

The Restricted Problem

One notes that if only two of the bodies have non-zero mass, the proposed method is strictly symplectic. The motions of the massive bodies are then integrated exactly (and are analytic functions of time), and for each massless body the Hamiltonian has the form

$$H = \frac{\mathbf{v}^2}{2} - \sum_j \frac{m_j}{|\mathbf{r} - \mathbf{r}_j(t)|} + P_t, \tag{6.93}$$

valid in the extended phase space where time is a coordinate. Because the momentum of time P_t satisfies the differential equation

$$\dot{P}_t = -\frac{\partial H}{\partial t}, \tag{6.94}$$

application of the symplectic generalized leapfrog, i.e. WH method, clearly leads to the algorithm described above. The proposed checking method is just a simple generalization of this.

6.3 Non-Canonical Forces

In this section the inclusion of non-canonical perturbations, such as drag forces, are considered in the framework of the δ-function formalism, as well as by using the device of formal Hamiltonization.

6.3.1 Analogue of the WH Method

Consider the perturbed two-body Hamiltonian

$$H = K + R(\mathbf{r}), \tag{6.95}$$

where K is the Keplerian part and R is a perturbation. The basic idea in the WH method is to replace this by the expression

$$H = K + \delta_h(t - \tfrac{1}{2}h)R(\mathbf{r}), \tag{6.96}$$

where h is the stepsize and δ_h is the delta-function fence.

If the system has non-canonical perturbations, it is not possible to maintain the Hamiltonian formalism as such. However, the differential equation following from the above Hamiltonian,

$$\ddot{\mathbf{r}} = \mathbf{F} + h\delta_h(t - \tfrac{1}{2}h)\mathbf{f}(\mathbf{r}) \tag{6.97}$$

(where **F** is the two-body acceleration and **f** is the perturbation), can immediately be generalized to

$$\ddot{\mathbf{r}} = \mathbf{F} + h\delta_h(t - \tfrac{1}{2}h)\mathbf{f}(\mathbf{r}, \mathbf{v}, t), \tag{6.98}$$

where $\mathbf{v} = \dot{\mathbf{r}}$ is the velocity. The solution for this differential equation is as follows.

1. The two-body motion is governed by the equations

$$\dot{\mathbf{r}} = \mathbf{v}, \quad \dot{\mathbf{v}} = \mathbf{F}, \tag{6.99}$$

   when the system moves between the point of action of the delta-function fence.
2. At the point of action of a delta-function, the solution must be obtained from

$$\dot{\mathbf{v}} = \mathbf{f}(\mathbf{r}, \mathbf{v}, t), \quad \dot{\mathbf{r}} = \mathbf{0}, \quad \dot{t} = 0, \tag{6.100}$$

   integrated over a time interval of length $= h$. This result is easily obtained, e.g. by replacing the δ-function by its approximation, a function with value $1/\varepsilon$ over a time interval of length ε, and letting $\varepsilon \to 0$. When the δ-function acts, the differential equations become

$$\dot{\mathbf{v}} = \frac{h}{\varepsilon}\mathbf{f}(\mathbf{r}, \mathbf{v}, t), \quad \dot{\mathbf{r}} = \mathbf{v}, \tag{6.101}$$

   in which the time advances over an interval of length ε. After adopting a new time variable $t_\delta = \varepsilon t/h$ one can write

$$\frac{d\mathbf{v}}{dt_\delta} = \mathbf{f}(\mathbf{r}, \mathbf{v}, t + \frac{\varepsilon}{h}t_\delta), \quad \frac{d\mathbf{r}}{dt_\delta} = \frac{\varepsilon}{h}\mathbf{v}, \tag{6.102}$$

   where the t_δ interval is h. In the limit, this reduces back to (6.100), where the dot is used as notation for the derivative.

Note, especially, that the result is *not* the same as using a velocity jump computed with the most recent values of the variables $\mathbf{r}, \mathbf{v}$ and t (Cordeiro et al., 1997), but the variation of $\mathbf{v}$ must be obtained by solving the differential equation (6.100) where $\mathbf{r}$ and t are considered constants. This implies the use of a non-symplectic composite method, instead of a symplectic one. The non-symplecticness is then in the physics of the system and cannot be avoided.

6.3.2 Formal Hamiltonian

A general method for considering non-conservative systems within the Hamiltonian formulation was proposed by Kamel (1971) and also by Choi and Tapley (1973). This involves impeding the n-dimensional non-conservative system in a $2n$-dimensional space where the system can be made formally Hamiltonian. Let us introduce the Hamiltonian

$$H = p_0 + \mathbf{P}_r \cdot \mathbf{v} + \mathbf{P}_v \cdot \mathbf{F} + \mathbf{P}_v \cdot \mathbf{f}(\mathbf{r}, \mathbf{v}, t), \tag{6.103}$$

where t, $\mathbf{r}$ and $\mathbf{v}$ are considered as coordinates and p_0, $\mathbf{P}_v$ and $\mathbf{P}_r$ their conjugate momenta. This formal Hamiltonian gives the equations of motion

$$\dot{\mathbf{v}} = \frac{\partial H}{\partial \mathbf{P}_v} = \mathbf{F} + \mathbf{f}, \quad \dot{\mathbf{r}} = \frac{\partial H}{\partial \mathbf{P}_r} = \mathbf{v}, \quad \dot{t} = \frac{\partial H}{\partial p_0} = 1, \tag{6.104}$$

which are obviously correct.

Now dividing the Hamiltonian into two parts,

$$H_0 = p_0 + \mathbf{P}_r \cdot \mathbf{v} + \mathbf{P}_v \cdot \mathbf{F}, \quad H_1 = \mathbf{P}_v \cdot \mathbf{f}(\mathbf{r}, \mathbf{v}, t), \tag{6.105}$$

and applying the generalized leapfrog principle produces the 'unperturbed step equations':

$$\dot{\mathbf{v}} = \frac{\partial H_0}{\partial \mathbf{P}_v} = \mathbf{F}, \quad \dot{\mathbf{r}} = \frac{\partial H_0}{\partial \mathbf{P}_r} = \mathbf{v}, \quad \dot{t} = \frac{\partial H_0}{\partial p_0} = 1. \tag{6.106}$$

From the perturbation part one obtains

$$\dot{\mathbf{v}} = \frac{\partial H_1}{\partial \mathbf{P}_v} = \mathbf{f}, \quad \dot{\mathbf{r}} = \frac{\partial H_1}{\partial \mathbf{P}_r} = 0, \quad \dot{t} = \frac{\partial H_1}{\partial p_0} = 0, \tag{6.107}$$

which are the same as above, while the rest of the equations (for the formal momenta) are not needed in practice.

More generally, by defining a formal Hamiltonian

$$H = \mathbf{P} \cdot \Phi(\mathbf{x}), \tag{6.108}$$

it is possible to apply the above to any flow

$$\dot{\mathbf{x}} = \Phi(\mathbf{x}). \tag{6.109}$$

If it is possible to write

$$\Phi(\mathbf{x}) = \mathbf{F}(\mathbf{x}) + \mathbf{f}(\mathbf{x}), \tag{6.110}$$

one may divide the Hamiltonian into two parts,

$$H_0 = \mathbf{P} \cdot \mathbf{F}, \quad H_1 = \mathbf{P} \cdot \mathbf{f}, \tag{6.111}$$

and subsequently apply the generalized leapfrog to this system (assuming both parts to be individually integrable).

However, the perturbation part is often not easily integrable analytically. If it is small, an approximate (numerical) solution is sufficient. The formal Hamiltonian formulation guides one in selecting the method. If the 'unperturbed' part of the derivatives, and possibly a major part of the perturbation, arises from a conservative Hamiltonian system, then a symplectic method is advisable. A simple choice for this is the implicit midpoint method (5.48), which can be efficient despite its implicit character, if the perturbation is sufficiently weak. Another reason for this choice is that it gives a time-reversible result if the system itself is time-reversible. Also, time-reversibility alone is normally sufficient for good energy conservation in long-term integrations (Hut, Makino and McMillan, 1995; Funato et al., 1996).

6.4 Alternative Formulations

6.4.1 Heliocentric Coordinates

Let $\mathbf{p}_k$ and $\mathbf{r}_k$ be the momenta and coordinates in an inertial system and let the masses be $m_0, m_1, \ldots, m_N$, where m_0 is the mass of the Sun and m_k, $k = 1, \ldots, N$, are the masses of the planets. To transform the Hamiltonian into heliocentric coordinates, the generating function

$$S = \mathbf{w}_0 \cdot \mathbf{x}_0 + \sum_{k=1}^{N} \mathbf{w}_k \cdot \mathbf{x}_k \tag{6.112}$$

may be used, where $\mathbf{x}_k = \mathbf{r}_k - \mathbf{r}_0$ are the heliocentric coordinates of the planets and $\mathbf{x}_0 = \sum_{k=1}^{N} m_k \mathbf{r}_k / m_{tot}$ is the centre of mass. The transformation formulae for the momenta are

$$\mathbf{p}_k = \mathbf{w}_k + \frac{m_k}{m_{tot}} \mathbf{w}_0, \quad \mathbf{p}_0 = -\sum_{k=1}^{N} \mathbf{w}_k + \mathbf{w}_0 \frac{m_0}{m_{tot}}. \tag{6.113}$$

The Hamiltonian takes the form

$$H = \frac{1}{2m_{tot}} \mathbf{w}_0^2 + \frac{1}{2m_0} \left(\sum_{k=1}^{N} \mathbf{w}_k \right)^2 + \sum_{k=1}^{N} \frac{1}{2m_k} \mathbf{w}_k^2 - \sum_{i<j} \frac{m_i m_j}{|\mathbf{x}_i - \mathbf{x}_j|}. \tag{6.114}$$

From the transformation formulae, it is evident that the new momentum $\mathbf{w}_0$ is the momentum of the centre of mass while the other momenta $\mathbf{w}_k$, $k = 1, \ldots, N$, are the momenta of the planets in the inertial centre-of-mass system. They are thus, in a sense, in a system different from the new heliocentric coordinates. Despite this the Hamiltonian can be divided into a main (H_0) and an interaction (H_1) part:

$$H_0 = \frac{1}{2m_{tot}}\mathbf{w}_0^2 + \sum_{k=1}^{N}\left(\frac{1}{2m_k}\mathbf{w}_k^2 - \frac{m_0 m_k}{|\mathbf{x}_k|}\right), \tag{6.115}$$

$$H_1 = \frac{1}{2m_0}\left(\sum_{k=1}^{N}\mathbf{w}_k\right)^2 - \sum_{0<i<j}\frac{m_i m_j}{|\mathbf{x}_i - \mathbf{x}_j|}, \tag{6.116}$$

both of which are individually integrable. The Keplerian part gives two-body motions defined by the second-order differential equations

$$\ddot{\mathbf{x}}_k = -m_0\frac{\mathbf{x}_k}{|\mathbf{x}_k|^3},$$

thus indicating that, in this basic approximation, all the planets move around the same body (m_0) like massless particles. The term containing $\mathbf{w}_0$ simply moves the centre of mass linearly and can normally be dropped because $\mathbf{x}_0$ is ignorable. The interaction part depends on the momenta only through the sum $\sum \mathbf{w}_k$, so that all the coordinates have the same derivatives. Consequently the differences, on which the potential and forces solely depend, remain constants. Similarly, the forces remain constants and thus must only be evaluated once. The result is that the momentum jumps

$$\delta\mathbf{w}_k = -\frac{h}{m_k}\frac{\partial H_1}{\partial \mathbf{x}_k} \tag{6.117}$$

occur as usual, but these are accompanied by position increments

$$\delta\mathbf{x}_k = h\frac{1}{m_0}\left(\sum_{k=1}^{N}\mathbf{w}_k\right), \tag{6.118}$$

which are all the same and thus correspond to the motion of the Sun.

The advantage of the heliocentric formulation is that the ordering of the bodies is immaterial (contrary to the use of the Jacobian coordinates) and the evaluation of the interaction terms is simpler because no conversions between coordinate systems are needed. The drawback is that the momenta (and corresponding Keplerian velocities) are referred to the centre of mass of the system, a situation that causes problems if some orbits come close to the Sun (and/or the centre of mass). In addition, the Keplerian approximation in general is not as good as in the Jacobian coordinates.

6.4.2 Close Approaches

Close approaches of planetary bodies are a major problem for symplectic integrators. It is necessary to shorten the integration step, so that the method is no longer symplectic (since the stepsize then depends on the behaviour of the system) and the good conservation properties of the method are lost. However, some special ways to handle the stepsize reduction have been developed. There are ways to artificially divide the potential into different parts and shorten the timestep in a symplectic way (Duncan, Levison and Lee, 1998).

An elegant way to handle the close encounters was published by Chambers (1999). This method is based on a manipulation of the potential by moving a part of the perturbation potential into the Keplerian portion. This is done in such a way that the perturbation remains small even in the case of a close encounter, and the strong interaction is integrated numerically, over a short interval of time, in connection with the Keplerian motions. Outside close encounters, this method runs just as fast as the normal WH method, but occasionally a short numerical integration must be done (to high accuracy). Chambers uses the heliocentric coordinates and rewrites the heliocentric system Hamiltonian in the form

$$
\begin{aligned}
H_0 = {} & \frac{1}{2m_{tot}}\mathbf{w}_0^2 + \sum_{k=1}^{N}\left(\frac{1}{2m_k}\mathbf{w}_k^2 - \frac{m_0 m_k}{|\mathbf{x}_k|}\right) \\
& - \sum_{0<i<j} \frac{m_i m_j}{x_{ij}}(1 - K(x_{ij}/b_{ij})),
\end{aligned}
\tag{6.119}
$$

$$
H_1 = \frac{1}{2m_0}\left(\sum_{k=1}^{N}\mathbf{w}_k\right)^2 - \sum_{0<i<j}\frac{m_i m_j}{x_{ij}} K(x_{ij}/b_{ij}), \tag{6.120}
$$

where $K(z)$ is such that $K(z) = 0$ when z is small and $K(z) = 1$ when z is large. The critical distances b_{ij} are chosen to be a few times the mutual Hill radius of the interacting pair. Thus, when the distance between two bodies is large, there is no modification in the potential. Only at short distances does the mutual interaction of the close pair (say $x_{\alpha\beta}$) move to the H_0 part of the Hamiltonian. Then the three-body system (m_0, m_α, m_β) must be numerically integrated over the duration of the encounter. The Hamiltonian for this triplet is

$$
H_{\alpha\beta} = \sum_{k=\alpha,\beta}\left(\frac{1}{2m_k}\mathbf{w}_k^2 - \frac{m_0 m_k}{|\mathbf{x}_k|}\right) - \frac{m_\alpha m_\beta}{x_{\alpha\beta}}(1 - K(x_{\alpha\beta}/b_{\alpha\beta})). \tag{6.121}
$$

As such encounters are rare, this should not affect significantly the efficiency of the integrator, even when the numerical integration is done to the highest possible accuracy. In this way the integrator remains completely symplectic and one does not expect accuracy degradation.

The function $K(z)$ is unspecified thus far. Chambers suggests a rational function that is simple and therefore rapid to evaluate.

While doing the computation the K modification is taken into account only when a pair is sufficiently close together. During the distance evaluations, one may check for any distance short enough that the modification must be done. Then it is simple to evaluate the formal two-body pericentre distance for such a pair to decide whether and when the three-body Hamiltonian (6.121) must be numerically integrated.

6.4.3 Individual Timesteps

Saha and Tremaine (1994) suggested a symplectic method which uses individual timesteps. A simplified way of discussing this method concisely is the following. The Hamiltonian is written in the form

$$H = \sum_{k=1}^{N} K_k(\mathbf{p}_k, \mathbf{r}_k) + R(\mathbf{r}_1, \ldots, \mathbf{r}_N), \tag{6.122}$$

where the K_k are Keplerian Hamiltonians and R is the perturbation. In a planetary system, the periods of the planets can vary by several orders of magnitude and thus it seems unnecessary to compute all the motions with the small stepsize determined by the smallest of the periods in the system. A way to write the Saha–Tremaine algorithm in terms of the δ-function fences is similar to the expression (6.51):

$$H = \sum_{k=1}^{N} h_k \delta_{h_k}(t) K_k(\mathbf{p}_k, \mathbf{r}_k) + R(\mathbf{r}_1, \ldots, \mathbf{r}_N), \tag{6.123}$$

where the δ_k are periodic δ-functions with periods h_k. The stepsizes are so chosen (h_k/h_{k-1} = integer) that it is possible to have a time-symmetrical, reversible algorithm. Thus every time a δ-function acts, the corresponding Keplerian Hamiltonian causes a Kepler-motion jump and the body moves to the next position. The interaction part R only moves the velocities and it is acting continually.

In a practical implementation, one can reduce the amount of computation considerably by taking into account that, for any body, the velocities must only be updated at certain times, essentially just before the body makes its Kepler jump.

For the three-body case (just two planets), the algorithm is fairly simple to understand. First the outermost planet makes its Kepler jump moving to the midpoint of its first step (so that now it is half a step ahead of the inner one). Then the inner planet is moved, taking shorter Kepler-motion steps, until it has proceeded to the end of the outer planet's first step (so that the outer body is half a step behind the inner one). At every 'stop' of the inner body, the velocity increments for both planets are evaluated and added to their velocities. Then the outer body is moved again, this time the timestep being of its full length so that it arrives at the halfway point of its second step, being again a half an (outer-body) step ahead. The rest of the computation is then just repetition.

For N planets the algorithm is just a generalization of the above. First the outermost planet executes its (half) Kepler jump; then the next one closer to the central body (Sun) is moved and the increments of the velocities of the two are calculated and added. In fact, at every 'stop' of a planet, the interactions of the just moved body and those on upper orbits are computed and the increments are added to the velocities.

Saha and Tremaine (1994) describe the algorithm in terms of a recursive procedure, which may be easily converted to (most) programming languages:

```
Procedure   TICK(i)
            Advance    K_i(τ_i/2) and I_i(τ_i)
            IF I > 1   LOOP τ_i/τ_{i-1} TIMES
                       CALL TICK(i-1)
                       END LOOP
            END IF
            Advance    K_i(τ_i/2)
END         TICK
```

Here $K_i(\tau)$ indicates Keplerian motion (with timestep $= \tau$) of body number i, while $I_i(\tau)$ signifies the interaction of body i with all the outer ones (and addition of the velocity increments, corresponding to timestep τ, to the velocities of the bodies in interaction). Once a computer program, working like the above procedure, is available, the entire computation consists simply of the call `TICK(N)`, which advances the entire system over one outermost planet step.

The above formulation is not precisely the most effective one, for it always splits the Kepler motions into two half-step advances, but this formulation is reasonably easy to comprehend.

According to Saha and Tremaine, the method gives, in practice, a speedup factor of about two in Solar System computations. While this may be significant in very long computations, it is not a massive advantage. The method also has the downside that it may produce stepsize resonances that ruin the accuracy.

6.4.4 Warm-Up

A major reason for position error in symplectic integration is the difference of frequencies (mean motions) between the true system and the discrete mapping model. Saha and Tremaine (1992) presented an idea, based on adiabatic invariants, to improve the long-term accuracy of symplectic integrators in planetary system computations. The idea is to use the fact that a very slow change of the Hamiltonian preserves the actions, which are adiabatic invariants, and which give the mean motions of the planets. The surrogate Hamiltonian (6.38) equals the correct one if the stepsize $h = 0$. Thus, if the stepsize is slowly changed, starting with a very small one and gradually increasing it to the final value (which is used for the bulk of the calculation), then the Hamiltonian changes slowly, conserving the actions. One can therefore expect an increased accuracy in longitudes resulting from improved values of the mean motions.

This procedure of slowly increasing the integration stepsize is called the *warm-up* of the integrator. To do this in practice, one may, for example, integrate back and forth around the starting position, slowly increasing the stepsize. From numerical tests, this procedure substantially increases the precision. A simple heuristic way to understand how warm-up works is as follows. Normally, symplectic integration has a periodic error. When the integration is started, the error wave is entered at an unpredictable phase and the error has a non-zero average (though constant) value. However, when the stepsize is initially very small, the error amplitude is small, and when the stepsize increases, the amplitude increases too, but in such a way that the mean value remains near zero.

Figure 6.5 demonstrates the effect of the warm-up in the case of an elliptical orbit of low eccentricity, integrated using the basic leapfrog. The integration started at pericentre. Starting at some other phase would have correspondingly changed the phase in the error curve. Therefore the mean energy error, and thus the phase error, depend on the starting phase. With the warm-up procedure one can make sure that the mean energy error remains considerably smaller than the amplitude of the error oscillation.

6.5 Near-Integrable Systems

Consider an almost integrable system which can be represented by a Hamiltonian K with a perturbation added to it. Let $\mathbf{w}$ and $\mathbf{r}$ be

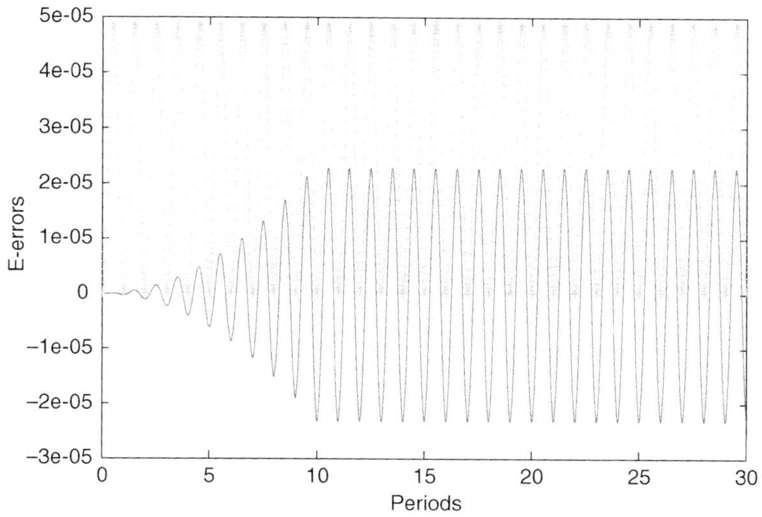

Fig. 6.5. Behaviour of the energy error with and without warm-up. The test orbit was a two-body orbit of eccentricity $e = 0.02$ and the method used was the basic leapfrog with a stepsize $h =$ period/64. Integration started at pericentre and, in the warm-up test, the stepsize increased linearly with time from a value $h/100$ to h in 10 periods. The warm-up curve starts from zero and has nearly zero average continuously. This shows the advantage of that method.

the momenta and coordinates of the system and t the time. The Hamiltonian of the perturbed system has the form

$$K(\mathbf{w}, \mathbf{r}) + \epsilon R(\mathbf{r}, t), \tag{6.124}$$

where ϵR is a small perturbation. In extended phase space, define new momenta $\mathbf{p}$ and coordinates $\mathbf{q}$ by

$$\mathbf{p} = (p_0, \mathbf{w}), \quad \mathbf{q} = (q_0, \mathbf{r}). \tag{6.125}$$

Using these, the Hamiltonian can be written as

$$H(\mathbf{p}, \mathbf{q}) = K(\mathbf{w}, \mathbf{r}) + p_0 + \epsilon R(\mathbf{r}, q_0), \tag{6.126}$$

which is equivalent to (6.124) but is formally independent of time.

The factor ϵ is included to indicate that only the first-order perturbations to the system will be considered. In this case, the symplectic method can be transformed into a conventional numerical integration of known functions of time, as will be shown in the next section.

Dividing the Hamiltonian as $H_0 = K + p_0$ and $H_1 = \epsilon R$, one can write for the two alternative WH leapfrog kernels the expressions

$$e^{(h/2)\widehat{H}_0} e^{h\widehat{H}_1} e^{(h/2)\widehat{H}_0}, \tag{6.127}$$

$$e^{(h/2)\widehat{H}_1} e^{h\widehat{H}_0} e^{(h/2)\widehat{H}_1}, \tag{6.128}$$

which can be generalized to create new, more complicated, kernels of the form

$$\prod_k e^{\alpha_k h \widehat{H}_0} e^{\beta_k h \widehat{H}_1}, \tag{6.129}$$

where α_k and β_k are constants. These have been proposed by many investigators (e.g. Yoshida, 1990; Koseleff, 1993; McLachlan, 1995; Chambers and Murison, 2000; Laskar and Robutel, 2001). In δ-function notation, this is equivalent to

$$H_\delta = H_0 + H_1 \sum_{k=1}^{n} W_k h \delta_h(t - x_k h), \tag{6.130}$$

where W_k and x_k are constants. Comparison with (6.129) gives equivalence when $\alpha_1 = x_1$, $\alpha_2 = x_2 - x_1$, $\ldots$, $\alpha_n = x_n - x_{n-1}$, $\alpha_{n+1} = 1 - x_n$, and $\beta_k = W_k$, $k = 1, \ldots, n$, $\beta_{n+1} = 0$. It is important to note that this formulation can be used only if the α are all positive because motions backwards cannot be coded into the δ-function representation.

6.5.1 First-Order Perturbations

It is not difficult to show that one can generate a symplectic method from any reasonable conventional integration scheme. Consider the splitting of (6.126) into

$$H_0 = K + p_0, \quad H_1 = \epsilon R. \tag{6.131}$$

As the methods (6.129) considered here are coordinate-independent, canonical transformations can be applied. A convenient transformation is $(\mathbf{w}, \mathbf{r}) \to (\mathbf{P}, \mathbf{Q})$, where $\mathbf{P}$ and $\mathbf{Q}$ are constants in the unperturbed motion defined by K; if K is the Keplerian Hamiltonian, then a possible set of variables $\mathbf{P}$, $\mathbf{Q}$ are the (parametric) initial values of the two-body motion (Section 4.3.1). Then the Hamiltonian may be written

$$H = p_0 + \epsilon R(\mathbf{q}(\mathbf{P}, \mathbf{Q}, q_0), q_0), \tag{6.132}$$

and the division into two parts takes the form

$$H_0 = p_0, \quad H_1 = \epsilon \widetilde{R}(\mathbf{P}, \mathbf{Q}, q_0), \tag{6.133}$$

where $\widetilde{R} = R(\mathbf{q}(\mathbf{Q}, \mathbf{P}, q_0), q_0)$. When a step forward is taken using the part H_0, we are simply updating time q_0, since the equations of motion reduce to the single equation

$$\dot{q}_0 = 1. \tag{6.134}$$

The perturbation part $H_1 = \epsilon\widetilde{R}$ leads to the update equations

$$\dot{P} = -\epsilon\frac{\partial\widetilde{R}}{\partial Q}(\mathbf{P},\mathbf{Q},q_0),$$
$$\dot{Q} = +\epsilon\frac{\partial\widetilde{R}}{\partial P}(\mathbf{P},\mathbf{Q},q_0),$$
$$\dot{q}_0 = +\epsilon\frac{\partial\widetilde{R}}{\partial p_0}(\mathbf{P},\mathbf{Q},q_0) = 0. \tag{6.135}$$

In general it would be difficult to solve these equations in this formulation. However, if the consideration is limited to the first-order perturbations in ϵ, then the initial values of P and Q can be used in the expressions on the right-hand side. The δ-function formulation (6.130) then gives for the changes of the elements $\mathbf{Q}$,

$$\Delta\mathbf{Q} = \sum_{k=1}^{n} W_k h\frac{\epsilon\partial\widetilde{R}}{\partial\mathbf{P}}(\mathbf{P}(0),\mathbf{Q}(0),q_0(0)+x_k h), \tag{6.136}$$

with a similar expression for the changes of $\Delta\mathbf{P}$. In this way, the problem has been reduced to the integration of a known function of time. One can thus conclude that *any reasonable conventional numerical integration formula generates a symplectic method* of the form (6.130).

Consider next a scheme that integrates a function $f(t)$ by evaluating the function at a set of n sample points $x_k h$ and combines them in some weighted sum:

$$\int_0^h f(t)dt = \sum_{k=1}^{n} W_k h f(x_k h) + O(h^{K+1}). \tag{6.137}$$

This method can be used to generate a symplectic method which will have an error, after a large number of steps, of $O(\epsilon h^K) + O(\epsilon^2 h^2)$. The original WH method (equation (6.127)) corresponds to the midpoint method, while the other possible leapfrog (equation (6.128)), where the roles of H_0 and H_1 are exchanged, corresponds to the trapezoidal rule.

6.5.2 Symplectic Correctors

The errors generated by the symplectic propagation of the perturbed Hamiltonian correspond to some error Hamiltonian. One can find a canonical transformation of variables that will eliminate the error Hamiltonian and thus greatly improve the accuracy of the integration. This transformation can be expressed in terms of a Lie operator that must be applied before each step, and an inverse transformation at the end of each step. This operator can be expressed as an exponential of some $\widehat{\mathbf{C}}$, which is called the symplectic corrector.

Wisdom, Holman and Touma (1996) derived the symplectic corrector for their leapfrog method. Instead of using the kernel (6.127) one includes the forward and inverse transformations, so that, for each step, updates use the kernel

$$e^{\widehat{\mathbf{C}}} e^{(h/2)\widehat{\mathbf{H}}_0} e^{h\widehat{\mathbf{H}}_1} e^{(h/2)\widehat{\mathbf{H}}_0} e^{-\widehat{\mathbf{C}}}. \tag{6.138}$$

When n steps of forward propagation are taken with this method, all the corrections cancel except right at the start and the end of the interval:

$$\left(e^{\widehat{\mathbf{C}}} e^{(h/2)\widehat{\mathbf{H}}_0} e^{h\widehat{\mathbf{H}}_1} e^{(h/2)\widehat{\mathbf{H}}_0} e^{-\widehat{\mathbf{C}}}\right)^n = e^{\widehat{\mathbf{C}}} e^{(h/2)\widehat{\mathbf{H}}_0} \left(e^{h\widehat{\mathbf{H}}_1} e^{h\widehat{\mathbf{H}}_0}\right)^n e^{-(h/2)\widehat{\mathbf{H}}_0} e^{-\widehat{\mathbf{C}}}. \tag{6.139}$$

Hence the use of a corrector does not significantly increase the computational effort if frequent output is not required.

Wisdom, Holman and Touma (1996) (WHT) showed that the leapfrog corrector could be computed to first order in ϵ in the operator form

$$C = G(\widehat{\mathbf{H}}_0)H_1, \tag{6.140}$$

where the function G is

$$G(Z) = -\sum_{j=0}^{\infty} B_j(x)\frac{(h)^j}{j!}Z^{j-1} = \frac{1}{Z} - \frac{he^{xhZ}}{(e^{hZ}-1)}. \tag{6.141}$$

The quantity $x = \mathrm{des}(t/h + 1/2)$, where des indicates decimal part, and $B_k(x)$ are the Bernoulli polynomials ($B_0 = 1$, $B_k(x) = k\int B_{k-1}(x)dx$, and $\int_0^1 B_k(x)dx = 0,\ k = 1, 2, \ldots$).

In addition to being a corrector, this expression also provides an interpolator, because the result can be evaluated at any point in time. For the midpoint scheme the increment is evaluated at $x = \frac{1}{2}$. In this case, the generating function G simplifies to

$$G(Z) = \frac{1}{Z} - \frac{he^{(h/2)Z}}{(e^{hZ}-1)}. \tag{6.142}$$

It is not difficult to show that this can be derived from the conventional numerical integration formalism. If the Hamiltonian (6.133) is used in the expression (6.140), then the operator $\widehat{\mathbf{H}}_0$ reduces to the time derivative $D \equiv \partial\partial/\partial\partial q_0$. Hence, for any numerical scheme the appropriate corrector can be formed from the relation

$$C = G(D)H_1. \tag{6.143}$$

Using this result, the generating function for the WHT corrector can be re-derived from the Euler–Maclaurin summation formula for the midpoint rule. Consider the error $\delta\sigma$ generated by integrating a function $f(t)$ over an interval nh by using the midpoint rule:

$$\delta\sigma = \int_0^{nh} f(t)dt - \sum_{k=0}^{n} hf((k+\tfrac{1}{2})h). \tag{6.144}$$

Next, using the differential operator D, write $f(t) = e^{tD}f(0)$, so that

$$\int_0^{nh} f(t)dt = \frac{e^{nhD}-1}{D}f(0). \tag{6.145}$$

This operator can be used to evaluate $f((k+\frac{1}{2})h)$ to show that the sum in (6.144) is

$$\sum_{k=0}^{n} hf((k+\tfrac{1}{2})h) = e^{(1/2)hD}\frac{e^{nhD}-1}{e^{hD}-1}f(0). \tag{6.146}$$

If one further notes that $f(nh) = e^{nhD}f(0)$ then the error is

$$\delta\sigma = \left(\frac{1}{D} - \frac{he^{(1/2)hD}}{e^{hD}-1}\right)[f(nh) - f(0)]. \tag{6.147}$$

This function is the same generating function as in (6.142) as a function of D.

6.5.3 Generating Functions

Having derived the WHT corrector, one can use the same procedure for finding correctors to first order in ϵ for other symplectic methods based upon conventional integration formulae. In general the function $f(t)$ can be integrated by evaluating it at steps $x_k h$ and combining these function values using weights W_k. One can then easily derive the generalized formula

$$\delta\sigma = \left(\frac{1}{D} - \sum_{k=1}^{n} W_k\frac{he^{x_k hD}}{e^{hD}-1}\right)(f(nh) - f(0)). \tag{6.148}$$

Hence the generating function of the symplectic corrector for the more general methods, such as (6.129) and (6.130), is

$$G(Z) = \frac{1}{Z} - \sum_{k=1}^{n} W_k\frac{he^{x_k hZ}}{e^{hZ}-1}. \tag{6.149}$$

Applying this in (6.140) provides the corrector operator. One may expand the generating function G in powers of h:

$$G(Z) = \frac{1}{Z}\left(1 - \sum_k W_k\right) - \sum_{j=1}^{\infty}\left(\sum_{k=1}^{n} W_k B_j(x_k)\right) h^j Z^{j-1}, \quad (6.150)$$

where the functions B_j are again the Bernoulli polynomials. As the integrals of the Bernoulli polynomials are zero, apart from B_0, it follows that if the W_k and x_k are the weights and nodes of a numerical integration formula, then the corrector has the same order in h as the error of the numerical integration formula in question. This result enables construction of higher-order symplectic integration schemes.

6.5.4 Modified Potential

The accuracy of a scheme of order K is, after a large number of steps, $O(\epsilon h^K)+O(\epsilon^2 h^2)$. We therefore only benefit from higher-order symplectic schemes if the order ϵ of the perturbation is very small.

In some useful cases, the accuracy of the symplectic scheme can be improved by removing the $O(\epsilon^2 h^2)$ error. In general, it is possible (e.g. Koseleff, 1993) to use, instead of (6.138), the kernel

$$e^{\widehat{\mathbf{C}}} e^{(h/2)\widehat{\mathbf{H}}_0} e^{h\widehat{\mathbf{H}}_1^{mod}} e^{(h/2)\widehat{\mathbf{H}}_0} e^{-\widehat{\mathbf{C}}}, \quad (6.151)$$

where a modified perturbation H_1^{mod} of the form

$$H_1^{mod} = H_1 + h^2 b\{\{H_0, H_1\}, H_1\} \quad (6.152)$$

is included. With some method-dependent factor b, this eliminates the error terms of order $O(\epsilon^2 h^2)$. If $H_0 = K$ is the Keplerian Hamiltonian (or, more generally, separable of the form $H_0 = T(\mathbf{p}) - U(\mathbf{q})$ where the kinetic energy T is quadratic in the momenta), then the above Poisson bracket depends only on the coordinates. Thus one actually has a modified potential. In the special case that $H_0 = K = \frac{1}{2}\mathbf{w}^2 - M/r$ and $H_1 = R(\mathbf{r}, t)$, the modified potential can be written as

$$\epsilon R^{mod} = \epsilon R(\mathbf{r}, t) + \tfrac{1}{2}\epsilon^2 h^2 \chi \left|\frac{\partial R}{\partial \mathbf{r}}\right|^2. \quad (6.153)$$

Use of this in place of the original perturbing function leaves a residual error of $O(\epsilon h^K) + O(\epsilon^2 h^4)$.

The perturbative acceleration resulting from this modified potential may further be approximated by

$$\mathbf{F}^{mod} = \mathbf{F}(\mathbf{r} + \chi h^2 \mathbf{F}(\mathbf{r}, t), t), \quad (6.154)$$

an expression that is not exactly correct because it does not generally have a potential, but the error is of order ϵ^3, and is quite often negligible. To complete this approach, parameter χ must be computed. Since one knows that such an expression exists, the parameter can be calculated for every particular method by examining a test case. In the next section examples of how to compute χ are given.

6.5.5 Derivation of Alternative Schemes

Spacing of Nodes

It has been shown (Mikkola and Palmer, 2001) that, by a suitable choice of an integration formula, from which to generate a symplectic integrator, and using the modified potential, a scheme can be generated in which the error term is $O(\epsilon h^K) + O(\epsilon^2 h^4)$. Suitable candidate integration schemes are now examined. The Gaussian integration formulae look promising as these exploit the spacing of perturbation evaluations to eliminate higher-order terms. The two-node Gaussian formulae can be used to generate a $K = 4$ scheme. In this case two perturbation evaluations per step, using the kernel

$$e^{\widehat{\mathbf{C}}} e^{\alpha_1 h \widehat{\mathbf{H}}_0} e^{h \widehat{\mathbf{H}}_1} e^{\alpha_2 h \widehat{\mathbf{H}}_0} e^{h \widehat{\mathbf{H}}_1} e^{\alpha_1 h \widehat{\mathbf{H}}_0} e^{-\widehat{\mathbf{C}}}, \tag{6.155}$$

are necessary. The spacing is not uniform, with $\alpha_1 = 1 - 1/\sqrt{3}$ and $\alpha_2 = 2/\sqrt{3}$. This scheme is slightly better than the one using Simpson's rule, which has a kernel

$$e^{\widehat{\mathbf{C}}} e^{(h/3)\widehat{\mathbf{H}}_1} e^{h \widehat{\mathbf{H}}_0} e^{(4h/3)\widehat{\mathbf{H}}_1} e^{h \widehat{\mathbf{H}}_0} e^{(h/3)\widehat{\mathbf{H}}_1} e^{-\widehat{\mathbf{C}}}. \tag{6.156}$$

For these higher-order schemes, the use of a corrector makes little difference to the performance. One can, however, use the corrector to estimate the accuracy of the scheme. For the two-point Gaussian formula one finds

$$C_G = +\frac{h^4}{270} \widehat{\mathbf{H}}_0^3 H_1 + \cdots, \tag{6.157}$$

which is slightly smaller than that for Simpson's rule,

$$C_S = -\frac{h^4}{180} \widehat{\mathbf{H}}_0^3 H_1 + \cdots. \tag{6.158}$$

Numerical experiments, however, have shown that normally the basic WH with leading-order corrector $C \approx (h^2/24)\widehat{\mathbf{H}}_0 H_1$ gives comparable or even better accuracy than these higher-order methods (assuming their correctors are neglected). This occurs because in the WH method

$$C = \frac{h^2}{24}\widehat{\mathbf{H}}_0 H_1 + \frac{7h^4}{5760}\widehat{\mathbf{H}}_0^3 H_1 + \cdots, \tag{6.159}$$

and thus the remaining corrector is smaller here than in the above methods.

The corrector can, however, be used to improve the order of the method without increasing the number of perturbation evaluations. To do this, the spacings at which to evaluate the perturbations are chosen suitably to remove terms in the generating function of the corrector:

$$\sum_k W_k = 1, \quad \sum_{k=1}^{n} W_k B_j(x_k) = 0, \tag{6.160}$$

for $j = 1$ and $j = 3, 4, 5, \ldots$, excluding $j = 2$. This means that the leading-order corrector $C \approx -(h^2/2!)\sum_k B_2(x_k)\widehat{\mathbf{H}}_0 H_1$ is not zero but is to be applied at output. This creates the possibility of obtaining higher precision using fewer perturbation evaluations. If a two-node formula is used, by symmetry $W_1 = W_2 = 1/2$. If one writes $x_1 = 1/2 - x$ and $x_2 = 1/2 + x$, then $B_4(1/2 - x) + B_4(1/2 + x) = 0$, or

$$x^4 - \frac{x^2}{2} + \frac{7}{240} = 0, \tag{6.161}$$

the solutions of which are $x = \pm\sqrt{1/4 \pm 1/\sqrt{30}}$. The relevant root is clearly

$$x = \sqrt{\frac{1}{4} - \frac{1}{\sqrt{30}}} \approx 0.25966481117961407141 80.$$

The step sequence in a long computation thus reads

$$2h(1/2 - x, 2x, 1 - 2x, 2x, 1 - 2x, \ldots, 2x, 1/2 - x),$$

where h has been replaced by $2h$ to make the mean step between perturbation evaluations the same as in the original WH method. The approximate numerical values in this are

$$(0.48067h, 1.03866h, 0.96134h, 1.03866h, 0.96134h, \ldots,$$
$$1.03866h, 0.48067h)$$

so that the only difference with the original WH is that the unperturbed stepsizes vary a little. When the leading-order corrector is applied,

$$C = h^2\left(\frac{2}{\sqrt{30}} - \frac{1}{3}\right)\widehat{\mathbf{H}}_0 H_1,$$

the remainder is of the order $O(\epsilon h^6) + O(\epsilon^2 h^2)$. In fact, the remainder term in the corrector is

$$\delta C = h^6 \frac{7\sqrt{30} - 45}{70875} \widehat{\mathbf{H}}_0^5 H_1 + \cdots . \tag{6.162}$$

Note: The three methods, i.e. the WH method, the two-point Gaussian and the alternating stepsize method, can all be unified into a formalism where the perturbation evaluations are taken at intervals of $(1 \pm q)h$. If we write $\alpha_1 = 1 - q$ and $\alpha_2 = 1 + q$, then the kernel for each step can be expressed as

$$e^{\widehat{\mathbf{C}}} e^{(\alpha_1/2)h\widehat{\mathbf{H}}_0} (e^{h\widehat{\mathbf{H}}_1} e^{\alpha_2 h\widehat{\mathbf{H}}_0} e^{h\widehat{\mathbf{H}}_1} e^{\alpha_1 h\widehat{\mathbf{H}}_0})^n e^{-(\alpha_1/2)h\widehat{\mathbf{H}}_0} e^{-\widehat{\mathbf{C}}}. \tag{6.163}$$

Hence, these methods can easily be implemented into a code in such a way that the value of the parameter q determines the choice of the method.

Parameters for the Corrector and Modified Potential

In many practically interesting cases, the Hamiltonian leads to equations of motion of the form

$$\ddot{\mathbf{q}} = \mathbf{N}(\mathbf{q}) + \mathbf{F}, \tag{6.164}$$

where $\mathbf{F} = \epsilon \partial R/\partial \mathbf{q}$. It is easy to show that, in this case, the leading-order correctors may be written as

$$\delta \mathbf{q} = -h^2 C_0 \mathbf{F}, \tag{6.165}$$

$$\delta \dot{\mathbf{q}} = h^2 C_0 \dot{\mathbf{F}}, \tag{6.166}$$

so that the corrector is reduced to a single parameter C_0. It is easy to show that

$$C_0 = -\frac{1}{24}(3q^2 + 6q - 1). \tag{6.167}$$

The coefficient χ, which appears in the modified potential, can also be determined. This is most easily done for the simple case of a free particle moving under the influence of a perturbation ϵR. This is determined by removing the leading error in the time-reversible integration of the Hamiltonian. One can then show that

$$\chi = (1 + 3q^2)/12. \tag{6.168}$$

This establishes relations for all the parameters for methods using a stepsize parameter q. If $q = 0$ the WH method is recovered, and if q is chosen as a root of equation (6.167) then the method reduces to the two-point Gaussian method. The higher-order method outlined above is obtained when $q = 0.03865924471845629$.

6.6 Time Transformations

If the nearly Keplerian orbits in the system under consideration are very eccentric, then the motion during one integration step is much faster near pericentre. In many cases this fact leads to inaccuracy, which can only be corrected by choosing initially a much shorter stepsize. Another possibility is the introduction of time transformations: e.g. introduce the time transformation from the physical time t to a new fictitious time s in the form

$$dt = g(\mathbf{q}, t)ds, \tag{6.169}$$

as well as a new Hamiltonian

$$\Gamma = g(\mathbf{q}, q_0)(H(\mathbf{p}, \mathbf{q}, q_0) + p_0), \tag{6.170}$$

where p_0 initially has the numerical value $p_0 = -H(0)$, and $q_0 = t$. Thus the time t is now a coordinate denoted by q_0, and p_0 is the corresponding momentum. Clearly Γ does not depend explicitly on the new independent variable s and thus $\Gamma = 0$ along the orbit, and the momentum of time has the value $p_0 = -H$ along the orbit. If it is possible to divide the Hamiltonian Γ into two individually integrable parts,

$$\Gamma = \Gamma_0 + \Gamma_1, \tag{6.171}$$

then the generalized leapfrog principle can be applied to this Hamiltonian. The constant stepsize is now an s-step and thus the physical timestep will vary according to (6.169). If, as an example, the perturbed two-body problem is considered and $g = r$ used, then the new independent variable is proportional to the eccentric anomaly, instead of time. This provides significant advantage in highly eccentric orbits (Mikkola, 1997; Rauch and Holman, 1999).

6.6.1 Perturbed Two-Body Problem

Physical Coordinates

If the Hamiltonian is written

$$H = \frac{1}{2}\mathbf{p}^2 - \frac{M}{r} + R(\mathbf{r}, t), \tag{6.172}$$

then the new time-transformed Hamiltonian Γ becomes

$$\Gamma = g\left(\frac{1}{2}\mathbf{p}^2 - \frac{M}{r} + R(\mathbf{r}, q_0) + p_0\right). \tag{6.173}$$

The obvious division into two parts is

$$\Gamma_0 = g\left(\frac{1}{2}\mathbf{p}^2 - \frac{M}{r} + p_0\right), \qquad \Gamma_1 = gR(\mathbf{r}, t). \tag{6.174}$$

If Γ_1 is assumed to be independent of all the momenta, then it will produce a momentum increment only and is thus easily integrable for any choice of $g = g(\mathbf{r}, q_0)$. However, Γ_0 does not simply reproduce the two-body motion one would obtain by dropping the perturbing function R from (6.172). In fact, most choices of g would make it difficult to integrate. Therefore the choice of g must be considered in more detail. To this end one first transforms back to the physical time by using the time-transformation function $1/g$. The function Γ_0 is not zero in numerical value, but it is a constant along the orbit over the timestep under consideration. Let its value be $\epsilon = \Gamma_0$. Then, in physical space,

$$\widetilde{H} = g^{-1}(\Gamma_0 - \epsilon) = \tfrac{1}{2}\mathbf{p}^2 - M/r + p_0 - \epsilon/g. \tag{6.175}$$

The result is thus a two-body Hamiltonian with the perturbation $-\epsilon/g$. If we, for example, take $g = r$, then

$$\widetilde{H} = g^{-1}(\Gamma_0 - \epsilon) = \tfrac{1}{2}\mathbf{p}^2 - (M + \epsilon)/r + p_0, \tag{6.176}$$

which is clearly the Hamiltonian of Keplerian motion, with a modified mass $M + \epsilon$. However, $g = r$ is not the only possibility leading to easily integrable systems. Consider more generally the transformation

$$g = \left(\frac{1}{r} + f(q_0)\right)^{-1} = \frac{r}{1 + rf(q_0)}, \tag{6.177}$$

which gives

$$\widetilde{H} = g^{-1}(\Gamma_0 - \epsilon) = \tfrac{1}{2}\mathbf{p}^2 - (M + \epsilon)/r + p_0 - \epsilon f(q_0). \tag{6.178}$$

Again the particle motion is modified-mass Keplerian, but now p_0 is variable because $\dot{p}_0 = -\partial\widetilde{H}/\partial q_0 = \epsilon \dot{f}(t)$, so that $p_0 - \epsilon f$ is a constant.

The time equation $t' = g$ must also be solved. Usually, the most convenient way to do this is to write $ds = dt/g$, which leads to

$$h = \int_0^t (1/r + f(q_0))dt = \int \frac{dt}{r} + \int f(t)dt. \tag{6.179}$$

Because r is now the radius vector of Keplerian motion, these integrals can be done, although only in implicit form. Let $\int dt/r = y$, then y satisfies the Stumpff's 'Hauptgleichung' form of Kepler's equation,

$$t = r_0 y + \mathbf{r}_0 \cdot \mathbf{v}_0 y^2 c_2(z) + M(1 - r_0/a) y^3 c_3(z), \tag{6.180}$$

where $\mathbf{v}_0$ is the velocity and a is the semi-major axis, while $c_2(z)$ and $c_3(z)$ are Stumpff functions with $z = My^2/a$. These are two equations with two unknowns t and y. Depending on $f(t)$, one more equation and unknown may appear as noted below.

Between the Keplerian steps one evaluates the momentum jumps by means of

$$\delta \mathbf{p} = -h \frac{\partial (gR)}{\partial \mathbf{r}}, \tag{6.181}$$

$$\delta p_0 = -h \frac{\partial (gR)}{\partial q_0}, \tag{6.182}$$

which are then to be added to the momenta before proceeding to the next Keplerian orbit step.

In the case of the eccentric restricted three-body problem, if the particle under consideration is in orbit with a much smaller semi-major axis than that of the perturber, a suitable choice for $f(t)$ is

$$f(t) = A/r_1(t), \tag{6.183}$$

where A is an adjustable constant and r_1 is the distance of the perturber from the central body. The advantage of this choice is that the physical timestep becomes smaller when the perturbing body is closer, thus presumably increasing precision. In this case, however, the time equation takes the form

$$h = y + Ay_1, \tag{6.184}$$

where $y_1 = \int dt/r_1$ and we have another Stumpff's 'Hauptgleichung' for the motion of the perturbing body.

The perturbing function R is in this case

$$R = -m_1 \left(\frac{1}{\Delta} - \frac{1}{r_1} - \frac{\mathbf{r} \cdot \mathbf{r}_1}{r_1^3} \right). \tag{6.185}$$

Here the term $-1/r_1$, which is a function only of time and thus does not affect the motion, is added to make the perturbation function smaller in numerical value. This also reduces the variation of p_0, hopefully making the numerical results more accurate.

KS-Transformation

The KS variables (e.g. Stiefel and Scheifele, 1971) are related to the physical ones by

$$\mathbf{r} = \widehat{\mathbf{Q}}\mathbf{Q}, \quad \mathbf{p} = \frac{\widehat{\mathbf{Q}}\mathbf{P}}{2\mathbf{Q}^2}, \quad r = \mathbf{Q}^2, \tag{6.186}$$

where $\widehat{\mathbf{Q}}$ is the KS matrix. With these and the time transformation $dt = r\,ds$, the resulting Hamiltonian is

$$\Gamma = \tfrac{1}{8}\mathbf{P}^2 - M + \mathbf{Q}^2 p_0 + \mathbf{Q}^2 R(\mathbf{Q}, q_0), \tag{6.187}$$

which has the obvious splitting

$$\Gamma_0 = \tfrac{1}{8}\mathbf{P}^2 - M + p_0\mathbf{Q}^2, \quad \Gamma_1 = \mathbf{Q}^2 R. \tag{6.188}$$

Clearly the main part Γ_0 represents a harmonic oscillator, and the momentum jumps are given by

$$\delta\mathbf{P} = -h\frac{\partial(\mathbf{Q}^2 R(\mathbf{Q}, q_0))}{\partial\mathbf{Q}}, \quad \delta p_0 = -h\frac{\partial(\mathbf{Q}^2 R(\mathbf{Q}, q_0))}{\partial q_0}. \tag{6.189}$$

To use the other time transformation $g = (1/r + f(t))^{-1}$, one must apply the additional time transformation $ds = g_1 ds_1$, where

$$g_1 = 1/(1 + rf(q_0)), \tag{6.190}$$

to obtain

$$\widetilde{\Gamma}_0 = g_1\Gamma_0, \quad \widetilde{\Gamma}_1 = g_1\Gamma_1. \tag{6.191}$$

Transforming $\widetilde{\Gamma}_0$ back to the s system, one obtains, as before,

$$\begin{aligned}\widetilde{\widetilde{\Gamma}}_0 &= g_1^{-1}(g_1\Gamma_0 - \epsilon) = \Gamma_0 - \epsilon/g_1 \\ &= \tfrac{1}{8}\mathbf{P}^2 - (M + \epsilon) + (p_0 - \epsilon f(t))\mathbf{Q}^2,\end{aligned} \tag{6.192}$$

containing the modified mass $M + \epsilon$, as well as the constant $p_0 - \epsilon f(t)$, which defines the frequency ω of the oscillator

$$\omega = \sqrt{\tfrac{1}{2}(p_0 - \epsilon f(t))}. \tag{6.193}$$

The relation between s and s_1 can be integrated to

$$s_1 = s + \int_0^{t(s)} f(t)dt, \tag{6.194}$$

where $t(s)$ is obtained from the 'Hauptgleichung' of the form (6.180).

It is of some interest to compare the physical-space treatment and KS treatment. Are there differences? It is clear that the two-body motions should be the same independent of treatment. Then there remains the question of whether the momentum jumps give the same (physical) outcome in both spaces. Taking into account the relation between the physical momentum $\mathbf{p}$ and the KS momentum $\mathbf{P}$, i.e.

$$\mathbf{p} = \frac{\widehat{\mathbf{Q}}\mathbf{P}}{2\mathbf{Q}^2}, \tag{6.195}$$

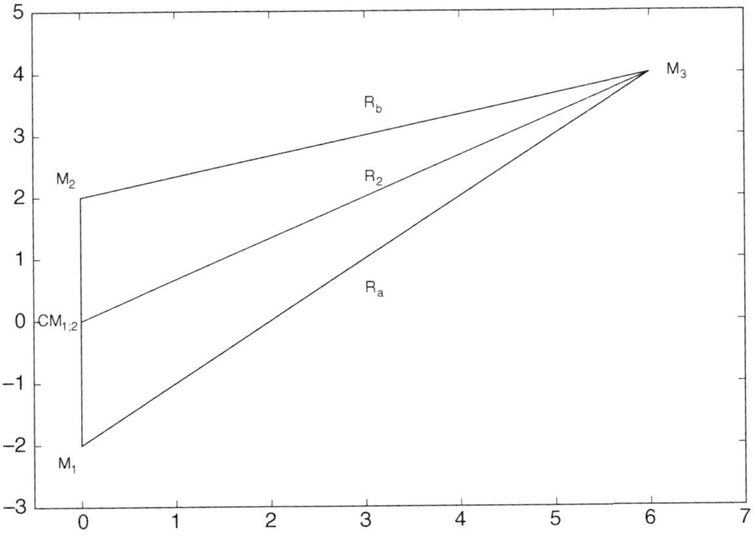

Fig. 6.6. Illustration of the three-body system and the notation used. The vector $\mathbf{R}_1$, from M_1 to M_2, is the relative position vector of the binary, while $\mathbf{R}_2$, from the centre of mass of the binary to M_3, defines the outer orbit.

the increment of physical momentum is

$$\delta\mathbf{p} = \frac{\widehat{\mathbf{Q}}\delta\mathbf{P}}{2\mathbf{Q}^2} = \frac{\widehat{\mathbf{Q}}\partial(gR)/\partial\mathbf{Q}}{2\mathbf{Q}^2} = \frac{2\widehat{\mathbf{Q}}\widehat{\mathbf{Q}}^t\partial(gR)/\partial\mathbf{r}}{2\mathbf{Q}^2} = \frac{\partial(gR)}{\partial\mathbf{r}}, \tag{6.196}$$

which is identical to the one obtained in the physical space. The basic reason is that the transformation between physical and KS momenta is linear with respect to the momenta. Thus not much difference is expected when using the regularized KS coordinates. However, for very eccentric orbits, the numerical stability may be affected because the physical momenta (velocities) have large variations while the KS momenta vary much less.

6.6.2 General Three-Body Problem

In terms of the Jacobian coordinates $\mathbf{r}_1$, $\mathbf{r}_2$ ($\mathbf{R}_1$ and $\mathbf{R}_2$ in Figure 6.6), and the corresponding momenta $\mathbf{p}_1$, $\mathbf{p}_2$, the Hamiltonian of a three-body system can be written in the form

$$H = K_1 + K_2 + R, \tag{6.197}$$

where

$$K_1 = \frac{1}{2\mu_1}\mathbf{p}_1^2 - \frac{m_1 m_2}{r_1}, \quad K_2 = \frac{1}{2\mu_2}\mathbf{p}_2^2 - \frac{m_3(m_1+m_2)}{r_2} \tag{6.198}$$

are the Keplerian Hamiltonians, while $\mu_1 = m_1 m_2/(m_1+m_2)$ and $\mu_2 = m_3(m_1+m_2)/(m_1+m_2+m_3)$. The perturbing function can be written

$$R = \frac{m_3(m_1 + m_2)}{r_2} - \frac{m_1 m_3}{r_a} - \frac{m_2 m_3}{r_b}. \tag{6.199}$$

As before, define

$$\Gamma = g(H + p_0), \tag{6.200}$$

where $p_0(0) = -H(0)$. Splitting Γ in the form

$$\Gamma_0 = g(K_1 + K_2 + p_0), \quad \Gamma_1 = gR, \tag{6.201}$$

and writing again for the numerical value $\epsilon = \Gamma_0$, as well as transforming back to physical time, one obtains

$$\widetilde{H}_0 = g^{-1}(\Gamma_0 - \epsilon) = K_1 + K_2 + p_0 - \epsilon/g. \tag{6.202}$$

As before, it is possible to make the choice

$$g^{-1} = A_1/r_1 + A_2/r_2, \tag{6.203}$$

where A_1 and A_2 are adjustable constants. The new Hamiltonian

$$H_0 = (K_1 - \epsilon A_1/r_1) + (K_2 - \epsilon A_2/r_2) \tag{6.204}$$

results in two Keplerian motions with modified masses. The momentum jumps are obtained by

$$\delta \mathbf{p} = -h \frac{\partial \Gamma_1}{\partial \mathbf{r}}, \tag{6.205}$$

$$\delta p_0 = -h \frac{\partial \Gamma_1}{\partial q_0}. \tag{6.206}$$

The time equation is obtained by integrating (6.203) with respect to the physical time

$$h = A_1 \int dt/r_1 + A_2 \int dt/r_2, \tag{6.207}$$

where both integrals over the inverse two-body distances satisfy their own 'Hauptgleichung'. The result is three equations,

$$h = A_1 h_1 + A_2 h_2, \tag{6.208}$$

$$t = r_1 h_1 + \mathbf{r}_1 \cdot \mathbf{v}_1 h_1^2 c_2(z_1) + \widetilde{M}_1(1 - r_1/a_1) h_1^3 c_3(z_1), \tag{6.209}$$

$$t = r_2 h_2 + \mathbf{r}_2 \cdot \mathbf{v}_2 h_2^2 c_2(z_2) + \widetilde{M}_2(1 - r_2/a_2) h_1^3 c_3(z_2), \tag{6.210}$$

for the three unknowns t, h_1 and h_2.

Use of Symplectic Corrector

The leading-order approximation for the corrector C is given by

$$C = \frac{h^2}{24}\{H_0, H_1\}, \tag{6.211}$$

from which the approximate corrections for coordinates and momenta are obtained by

$$\delta \mathbf{r} = \frac{\partial W}{\partial \mathbf{p}}, \quad \delta \mathbf{p} = -\frac{\partial W}{\partial \mathbf{r}}. \tag{6.212}$$

One should note that this formulation is also valid for the time-transformed system, although the derivations presented have concentrated on the Newtonian case.

To find the initial conditions for the actual integration one must solve iteratively the equations

$$\mathbf{r} = \mathbf{r}_{true} - \frac{\partial C(\mathbf{r}, \mathbf{p})}{\partial \mathbf{p}}, \quad \mathbf{p} = \mathbf{p}_{true} + \frac{\partial C(\mathbf{r}, \mathbf{p})}{\partial \mathbf{r}}. \tag{6.213}$$

This iteration must be done only once before starting the calculation and thus the effort required is negligible.

Three-Body Algorithm

1. Before starting the bulk of the integration, the total energy $E = -p_0$ must be calculated, followed by computation of the modified initial conditions using the symplectic corrector (6.212). This can be done by solving iteratively the equations (6.213).
2. When the parametric initial conditions are obtained, one proceeds a half Keplerian step and advances both orbits by the f–g algorithm. Then the momentum increments are evaluated as gradients of Γ_1, converted to velocity jumps, and added to the velocities.
3. Subsequently the algorithm proceeds with full Keplerian step(s) followed by velocity jumps.
4. Finally, when output is required, a half Keplerian step is carried out, followed by the application of the symplectic corrector (6.212).

6.6.3 Generalized Time Transformations

The time transformation $g = r$ is essentially equivalent to the use of the eccentric anomaly as the independent variable. The alternative $g = r^2$ corresponds to the true anomaly, while if $g = 1$ (i.e. no transformation) one can say that the independent variable is the mean anomaly.

It is possible to parametrize the time transformation so as to include all the alternatives, i.e. mean, eccentric and true anomaly, as well as any linear combination in one expression:

$$ds = dt/g(r), \tag{6.214}$$

$$g(r) = 1/(B_0 + B_1/r + B_2/r^2), \tag{6.215}$$

where B_0, B_1, B_2 are adjustable parameters (constants). Let us consider the Hamiltonian $H = K + R$, where $K = \mathbf{p}^2/2 - m/r$ is the Keplerian part and $R = R(\mathbf{r}, t)$ is the perturbing potential. Following Mikkola (1997) we apply the Poincaré transformation to obtain the time-transformed Hamiltonian Γ in the form

$$\Gamma = g(r)(K + p_0 + R) = g(r)(K + p_0) + g(r)R(\mathbf{r}, t). \tag{6.216}$$

As before, transform back to the physical time system using, once again, the Poincaré transformation. The result is

$$H_0 = \frac{1}{g(r)}[g(r)(K + p_0) - \gamma_0] = K + p_0 - \frac{\gamma_0}{g(r)}, \tag{6.217}$$

where γ_0 is the constant (over one step) numerical value of Γ_0 computed at the beginning of the step. When the suggested expression for the time-transformation function $g(r)$ is substituted, there results

$$\begin{aligned} H_0 &= \mathbf{p}^2/2 - m/r + p_0 - \gamma_0(B_0 + B_2/r + B_2/r^2) \\ &= \mathbf{p}^2/2 - (m + \gamma_0 B_1)/r - \gamma_0 B_2/r^2 + p_0 - \gamma_0 B_0. \end{aligned} \tag{6.218}$$

Note that if $B_2 = 0$, this has the form of a two-body Hamiltonian with a modified mass. However, when $B_2 \neq 0$ a small extra term with $1/r^2$ dependence appears. If the fact that the motion is planar and transformation to polar coordinates (r, θ) are used, the above Hamiltonian can be written as

$$H_0 = \tfrac{1}{2}(p_r^2 + (p_\theta^2 - 2\gamma_0 B_2)/r^2) - \widetilde{m}/r, \tag{6.219}$$

where the unnecessary constant term has been dropped and $\widetilde{m} = m + \gamma_0 B_1$ is the modified mass. The generating function

$$S = \psi p_\psi = \psi p_\theta \sqrt{1 - 2\gamma_0 B_2/p_\theta^2}$$

gives the canonical transformation

$$p_\psi^2 = p_\theta^2 - 2\gamma_0 B_2,$$

$$\psi = \theta / \sqrt{1 + 2\gamma_0 B_2/p_\psi^2}.$$

After this, the Hamiltonian has once again the two-body form

$$H_0 = \tfrac{1}{2}(p_r^2 + p_\psi^2/r^2) - \widetilde{m}/r. \tag{6.220}$$

The motion cannot be expressed in terms of the usual two-body formulae, but a fairly simple formulation is available.

Adopting $\mathbf{r}_0$, $\mathbf{v}_0$ and t_0 to be the values of position vector, velocity vector and time at the beginning of a Γ_0 step, then the equations below can be used to obtain new values of these quantities at the end of the step, which has the length h_s in the independent variable. Note that the variable p_0 does not change during this step. The special function at_1 is defined in (6.248), and the c_ν are the Stumpff functions.

$$g = 1/(B_0 + B_1/r_0 + B_2/r_0^2), \tag{6.221}$$

$$\gamma_0 = (\mathbf{v}_0^2/2 - m/r_0 + p_0)g, \tag{6.222}$$

$$\widetilde{m} = m + \gamma_0 B_1, \tag{6.223}$$

$$\epsilon = \gamma_0 B_2, \tag{6.224}$$

$$\mathbf{p}_\theta = \mathbf{r}_0 \times \mathbf{v}_0, \tag{6.225}$$

$$p_\theta^2 = |\mathbf{p}_\theta|^2, \tag{6.226}$$

$$\eta = \mathbf{r}_0 \cdot \mathbf{v}_0, \tag{6.227}$$

$$\dot{r}_0 = \eta/r_0, \tag{6.228}$$

$$p_\psi^2 = p_\theta^2 - 2\epsilon, \tag{6.229}$$

$$w^2 = \dot{r}_0^2 + p_\psi^2/r_0^2, \tag{6.230}$$

$$\beta = 2\widetilde{m}/r_0 - w^2, \tag{6.231}$$

$$\zeta = \widetilde{m} - r_0\beta, \tag{6.232}$$

$$h = h_s/(B_0 r_0 + B_1 + B_2/r_0). \tag{6.233}$$

Beginning of iteration loop (for h)

$$G_\nu = h^\nu c_\nu(\beta h^2), \tag{6.234}$$

$$r = r_0 + \eta G_1 + \zeta G_2, \tag{6.235}$$

$$\tilde{f} = 1 - \widetilde{m} G_2/r_0, \tag{6.236}$$

$$\tilde{g} = r_0 G_1 + \eta G_2, \tag{6.237}$$

$$\xi = \tilde{g}/(r_0^2\tilde{f} + \eta\tilde{g}), \tag{6.238}$$

$$y_\psi = \mathrm{atan}(p_\psi \xi)/p_\psi = \xi\, \mathrm{at}_1(p_\psi^2 \xi^2), \tag{6.239}$$

$$\Delta t = r_0 h + \eta G_2 + \zeta G_3, \tag{6.240}$$

$$d = B_0 r + B_1 + B_2/r, \tag{6.241}$$

$$h = h - (B_0 \Delta t + B_1 h + B_2 y_\psi - h_s)/d. \tag{6.242}$$

End of iteration loop (exit at convergence)

$$\dot{r} = (\eta G_0 + \zeta G_1)/r, \tag{6.243}$$

$$\theta^2 = p_\theta^2 y_\psi^2, \tag{6.244}$$

$$\mathbf{r} = \frac{r}{r_0}(c_0(\theta^2)\mathbf{r}_0 + y_\psi c_1(\theta^2)\mathbf{p}_\theta \times \mathbf{r}_0), \tag{6.245}$$

$$\mathbf{v} = \frac{\dot{r}}{r}\mathbf{r} + \frac{1}{r^2}\mathbf{p}_\theta \times \mathbf{r}, \tag{6.246}$$

$$t = t_0 + \Delta t. \tag{6.247}$$

Here the special function at_1 is defined by

$$\begin{aligned}\mathrm{at}_1(z) &= \frac{\mathrm{atan}(\sqrt{z})}{\sqrt{z}}\\ &= \frac{1}{2}\log\left(\frac{1+\sqrt{-z}}{1-\sqrt{-z}}\right)\Big/\sqrt{-z}\\ &= 1 - z/3 + z^2/5 - z^3/7 + z^4/9 - z^5/11 + - \cdots .\end{aligned} \tag{6.248}$$

Finally, what remains to be explained here is the advancement of the system under the perturbation part $g(r)R(\mathbf{r},t)$ of the Hamiltonian. As this depends only on coordinates (time is also a coordinate now), the motion is linear in momentum space and the coordinates remain constant. We thus have the momentum increments

$$\delta\mathbf{v} = -h_s\frac{\partial(g(r)R(\mathbf{r},t))}{\partial\mathbf{r}}, \quad \delta p_0 = -h_s\frac{\partial(g(r)R(\mathbf{r},t))}{\partial t}, \tag{6.249}$$

which are to be added to the most recent values of $\mathbf{v}$ and p_0.

6.7 The Logarithmic Hamiltonian

In methods of regularization (e.g. Mikkola and Aarseth, 1993), it is possible to use a time-transformation function inversely proportional to the gravitational potential.

In general, the simplest way to obtain explicit composite methods for computing the motion under a Hamiltonian is to write the Hamiltonian in a separated form in which the parts contain canonical variables of only one or the other category. Thus write

$$H = T(\mathbf{p}) - U(\mathbf{r},t), \tag{6.250}$$

where $T(\mathbf{p})$ is the kinetic energy and $U(\mathbf{r},t)$ is the potential, assumed positive. If the time transformation

$$ds = U dt \tag{6.251}$$

is applied, where s is the new independent variable, defined by this differential equation, then the time-transformed Hamiltonian may be written

$$\Gamma = (T - U + p_0)/U. \tag{6.252}$$

Here $p_0 = -E_0$ is the negative of the energy of the system, calculated from initial values. Note that p_0 is also the conjugate momentum of time, which is now a coordinate. This Hamiltonian is not separable. However, $\Gamma = 0$ along the correct orbit. Thus it is possible to use the function

$$\Lambda = \log(1 + \Gamma) \tag{6.253}$$

as the new Hamiltonian. Clearly then

$$\mathbf{p}' = -\frac{\partial \Lambda}{\partial \mathbf{r}} = -\frac{1}{1+\Gamma}\frac{\partial \Gamma}{\partial \mathbf{r}}, \tag{6.254}$$

$$p_0' = -\frac{\partial \Lambda}{\partial t} = -\frac{1}{1+\Gamma}\frac{\partial \Gamma}{\partial t}, \tag{6.255}$$

$$\mathbf{r}' = \frac{\partial \Lambda}{\partial \mathbf{p}} = \frac{1}{1+\Gamma}\frac{\partial \Gamma}{\partial \mathbf{p}}, \tag{6.256}$$

$$t' = \frac{\partial \Lambda}{\partial p_0} = \frac{1}{1+\Gamma}\frac{\partial \Gamma}{\partial p_0}, \tag{6.257}$$

give correct equations of motion for $\mathbf{p}$, p_0, $\mathbf{r}$ and t. From the above expression for Γ,

$$1 + \Gamma = (T + p_0)/U, \tag{6.258}$$

the logarithm of which is the new Hamiltonian,

$$\Lambda = \log(T + p_0) - \log(U), \tag{6.259}$$

clearly in separable form. Thus it is possible to apply explicit composite methods to solve the corresponding equations of motion. An effect following from the application of the logarithm is that now $dt = ds/(T + p_0)$, instead of the original one. Such a change would not have occurred had we chosen initially $T + p_0$ as the time-transforming function and used $\Lambda = -\log(1 - \Gamma)$, instead of (6.253). The final result (6.259), however, remains the same.

6.7.1 The Exact Kepler Leapfrog

The logarithmic Hamiltonian for the two-body problem is

$$\Lambda_K = \ln\left(\frac{\mathbf{p}^2}{2} + p_0\right) + \ln(r) - \ln(M). \tag{6.260}$$

This gives the time-transformed equations of motion

$$\frac{dt}{ds} = \frac{1}{\mathbf{p}^2/2 + p_0}, \quad \frac{d\mathbf{r}}{ds} = \frac{\mathbf{p}}{\mathbf{p}^2/2 + p_0}, \quad \frac{d\mathbf{p}}{ds} = -\frac{\mathbf{r}}{r^2}, \quad \frac{dp_0}{ds} = 0, \tag{6.261}$$

which can be solved exactly using the leapfrog algorithm, except for a time error (Mikkola and Tanikawa, 1999a). However, Preto and Tremaine (1999) obtained an expression for the error of the time as well as for the precise meaning of the leapfrog stepsize. After a minor rewriting of the Preto and Tremaine (1999) expressions, the exact Kepler leapfrog algorithm results. Start with the values $\mathbf{p}_a$, $\mathbf{r}_a$, t_a at the beginning of a step of length h_s, and calculate the steplength for the leapfrog by

$$h = \frac{m}{\sqrt{p_0/2}} \tan\left(\frac{h_s}{m}\sqrt{\frac{p_0}{2}}\right), \tag{6.262}$$

where

$$m = r_a\left(\frac{\mathbf{p}_a^2}{2} + p_0\right) \tag{6.263}$$

is an effective mass. (This is used instead of the true mass M, because mass does not appear explicitly in the equations of motion and thus it enters the formulation only via the value of p_0.) Proceed with

$$\mathbf{r}_{1/2} = \mathbf{r}_a + \frac{h}{2}\mathbf{p}_a/(\mathbf{p}_a^2/2 + p_0), \tag{6.264}$$

$$\mathbf{p}_b = \mathbf{p}_a - h\,\mathbf{r}_{1/2}/r_{1/2}^2, \tag{6.265}$$

$$\mathbf{r}_b = \mathbf{r}_{1/2} + \frac{h}{2}\mathbf{p}_b/(\mathbf{p}_b^2/2 + p_0), \tag{6.266}$$

$$t_b = t_a + \frac{h}{2}\left(\frac{1}{\mathbf{p}_a^2/2 + p_0} + \frac{1}{\mathbf{p}_b^2/2 + p_0}\right) + \tau(h, m, P_t), \tag{6.267}$$

to obtain the values $\mathbf{p}_b$, $\mathbf{r}_b$, t_b at the end of the step. Here the time correction is $\tau = -h^3/(12m^2) + O(h^5)$, or precisely

$$\tau = -\frac{h_s^3}{4m^2}\frac{\tan(\sqrt{\zeta}) - \sqrt{\zeta})}{\sqrt{\zeta}^3}, \tag{6.268}$$

where

$$\zeta = \frac{h_s^2}{m^2}\frac{p_0}{2}. \tag{6.269}$$

In passing, one notes that the expression for h can also be written

$$h = (m/\sqrt{p_0/2})\tan(u/2), \tag{6.270}$$

where $u = E_b - E_a$ is the increment of the eccentric anomaly. Often it is convenient to use, instead of $s = m \int^t dt/r$, the variable $X = \int^t dt/r = s/m$ (because this is usually used in the f–g formulation). In terms of this, the leapfrog stepsize parameter can be written as

$$h = \frac{m}{\sqrt{p_0/2}} \tan(X\sqrt{p_0/2}) = mX \,\mathrm{tg}_1(X^2 p_0/2), \tag{6.271}$$

where the latter expression is formally independent of the orbit type (i.e. it does not contain the square root of p_0, so it is also a real expression for unbound orbits, which have $p_0 < 0$). The function $\mathrm{tg}_1(\zeta)$ is

$$\mathrm{tg}_1(\zeta) = \frac{\tan(\sqrt{\zeta})}{\sqrt{\zeta}} = 1 + \frac{\zeta}{3} + \frac{2\zeta^2}{15} + \frac{17\zeta^3}{315} + \cdots . \tag{6.272}$$

For this, as well as for the function

$$\mathrm{tg}_3(\zeta) = \frac{\tan(\sqrt{\zeta}) - \sqrt{\zeta}}{\sqrt{\zeta}^3} = \frac{1}{3} + \frac{2\zeta}{15} + \frac{17\zeta^2}{315} + \frac{62\zeta^3}{2835} + \cdots \tag{6.273}$$

needed in the expression (6.268) for τ, one can produce useful Padé approximants for easy evaluation (the notation tg_3 comes from $\tan(x) = x + x^3\,\mathrm{tg}_3(x^2)$). Using the Stumpff c-functions, it is possible to express the tg functions as

$$\mathrm{tg}_1(\zeta) = c_1(\zeta)/c_0(\zeta), \quad tg_3(\zeta) = (c_2(\zeta) - c_3(\zeta))/c_0(\zeta), \tag{6.274}$$

which is another convenient way to evaluate these functions.

Finally, for the unperturbed Kepler problem, the effective mass equals the physical mass, i.e. $m = M$, but in the case of perturbations this is no longer true because p_0 is then determined by the total Hamiltonian. Also it is worth mentioning that adding the time correction τ can be done without compromising the symplecticness of the method. This can be seen by adding, formally, the term $\int^{p_0} \tau dp_0/h$ to the logarithmic Hamiltonian.

A Toycode for Two-Body Orbit

```
implicit real*8 (a-h,m,o-z)
real*8 x(3),v(3)
twopi=8*atan(1.d0)
read(5,*)x,v,h,Ymx,M! read coord, vels,
step, max time,mass
t=0
r=sqrt(x(1)**2+x(2)**2+x(3)**2)
vv=v(1)**2+v(2)**2+v(3)**2
Pt=-vv/2+M/r   !this the binding energy
```

```
1       continue
        call xm(x,t,v,h/2,Pt) !x-motion
        call vj(x,v,h)        !v-jump
        call xm(x,t,v,h/2,Pt) !x-motion
        t=t-h**3/12/(M)**2! +O(h^{5})! time correction
        r=sqrt(x(1)**2+x(2)**2+x(3)**2)
        vv=v(1)**2+v(2)**2+v(3)**2
        hlog=log((vv/2+Pt)*r/M) ! logH
        write(6,*)t,' 0 0 0 ',x,hlog !~error
        if(y.lt.Ymx)goto 1
        end
        subroutine xm(x,t,v,h,Pt)
        implicit real*8 (a-h,m,o-z)
        real*8 x(3),v(3)
        dt=h/((v(1)**2+v(2)**2+v(3)**2)/2+Pt)
        do k=1,3
        x(k)=x(k)+dt*v(k)
        end do
        t=t+dt
        return
        end
        subroutine vj(x,v,h)
        implicit real*8 (a-h,m,o-z)
        real*8 x(3),v(3)
        dtc=h/(x(1)**2+x(2)**2+x(3)**2)
        do k=1,3
        v(k)=v(k)-dtc*x(k)
        end do
        return
        end
```

6.8 Functional Time Transformation

Preto and Tremaine (1999) noted the more general possibility that the logarithm can be replaced by any function $f(z)$ (which has a positive derivative $f'(z) > 0$),

$$\Lambda = f(T + p_0) - f(U). \tag{6.275}$$

The demonstration of the equivalence of this and the original Hamiltonian is very similar to the one with the logarithmic Hamiltonian and is omitted here. In this case, the time transformation is

$$\frac{dt}{ds} = f'(T + p_0), \tag{6.276}$$

which, along the correct orbit, is also $dt/ds = f'(U)$.

The Hamiltonian manipulation and the resulting time transformation can be generalized in several ways. Instead of dividing the Hamiltonian in such a way that separates the momenta and coordinates, one may use any other division $H = H_0 - R$ and a function $f(z)$ to obtain

$$\Lambda = f(H_0 + p_0) - f(R). \tag{6.277}$$

Furthermore one may use some other function $f(z, \mathbf{r}, \mathbf{p})$ and write $\Lambda = f(H_0 + p_0, \mathbf{r}, \mathbf{p}) - f(R, \mathbf{r}, \mathbf{p})$, which may be shown to be equivalent to the original Hamiltonian. This can be done in much the same way as before (but see below).

The particular case that we will consider here is somewhat simpler than the one above, namely

$$\Lambda = f(g(r)[K + p_0]) - f(g(r)R(t, \mathbf{r})), \tag{6.278}$$

where K is a Kepler Hamiltonian (but could in principle be any part of the original Hamiltonian) and R is the perturbing function. The separation into the Keplerian and perturbation parts is, however, not necessarily the normal one. One may, for example, split the $1/r$ potential in two parts, placing one part into the perturbing function. In some problems, this turns out to be a useful alternative, and will be discussed in more detail later.

First, a simple demonstration that Λ can be used as the Hamiltonian is provided. Following Preto and Tremaine (1999), we introduce a new time-transformation function $\tilde{g}$ defined by

$$\tilde{g}(p_0, \mathbf{p}, t, \mathbf{r}) = g\frac{f(z_0) - f(z_1)}{z_0 - z_1}, \quad z_0 = g(r)(K + p_0), \;\; z_1 = g(r)R(t, \mathbf{r}), \tag{6.279}$$

where a difference quotient $(f(z_0) - f(z_1))/(z_0 - z_1)$ resides, which reduces to the derivative $f'(z)$ in the case $z_0 \to z_1 \to z$. Thus the fact that, in the exact orbit, $z_0 = z_1$, does not imply any singularity of $\tilde{g}$. When the time-transformation function $\tilde{g}$ is used by multiplying the homogeneous Hamiltonian $K + p_0 - R$ by it, there results $\tilde{g}(p_0, \mathbf{p}, t, \mathbf{r})(K + p_0 - R) = \Lambda$. Therefore Λ can be used as the Hamiltonian in terms of our earlier arguments.

Let us now divide Λ into the two parts

$$\Lambda_0 = f(g(r)[K + p_0]), \quad \Lambda_1 = -f(g(r)R(t, \mathbf{r})), \tag{6.280}$$

and consider the motion under the Hamiltonian Λ_0. Writing $\Gamma_0 = g(r)(K + p_0)$, the equation of motion for the time is

$$\frac{dt}{ds} = f'(\Gamma_0)g(r). \tag{6.281}$$

From this, or by considering Hamilton's equation of motion, it is easy to see that the only difference from what we would obtain, for any of the dependent variables, using Λ_0 versus the basic Poincaré-transformed Hamiltonian Γ_0, is just the additional factor $f'(\Gamma_0)$. On the other hand, Γ_0 is constant over the integration step under consideration, so that the solution for the Λ_0 problem is just the same as that for Γ_0, save for the constant scaling factor in the (physical) timestep corresponding to the constant steplength in the new independent variable. Here it is also important to realize that $\Gamma_0 \approx g(r)R$ so that the time transformation can be considered to be

$$\frac{dt}{ds} \approx f'(g(r)R)g(r). \tag{6.282}$$

This is a crucial result, as it will allow the physical timestep to adjust in response to both the central potential and the perturbation. However, the details depend on our choice of the functions f and R (how much, if any, of the $1/r$ potential has been moved to the perturbing function).

The various consequences of this are most simply demonstrated by considering some specific examples, such as the problem of two fixed centres and the restricted three-body problem.

6.8.1 Some Special Cases

Two Fixed Centres

The problem of two fixed centres has the Hamiltonian

$$H = \tfrac{1}{2}\mathbf{p}^2 - m_1/r_1 - m_2/r_2, \tag{6.283}$$

where $\mathbf{r}_k = \mathbf{r} - \mathbf{x}_k$ and $\mathbf{x}_1$ and $\mathbf{x}_2$ are the positions of the two attracting centres. Consider now the case when $m_2 \ll m_1$, which has some formal resemblance to the problem of motions in a planetary system. Write $M = m_1$, $m = m_2$ and

$$H = \tfrac{1}{2}\mathbf{p}^2 - M/r - m/\Delta, \tag{6.284}$$

where $\mathbf{x}_1 = \mathbf{0}$ and thus $\Delta = |\mathbf{x}_1 - \mathbf{r}|$. At first we use $g(r) = 1$, and thus take, for the functional Hamiltonian, the form

$$\Lambda = f(\underbrace{\tfrac{1}{2}\mathbf{p}^2 - (M - \widetilde{m})/r + p_0}_{\Gamma_0}) - f(\underbrace{\widetilde{m}/r + m/\Delta}_{R}), \tag{6.285}$$

where $\widetilde{m}$ is an adjustable constant, which splits the Kepler potential into two parts, one of which we consider part of Γ_0, the other part of R. For the special case $f(z) = \ln(z)$, the time transformation is

$$\frac{dt}{ds} = 1/\Gamma_0 \approx 1/R = 1/(\widetilde{m}/r + m/\Delta) = \frac{r\Delta}{\widetilde{m}\Delta + mr}, \tag{6.286}$$

which ensures that the physical timestep becomes small as either r or Δ goes to zero. If, on the other hand, $g(r) = r$ and $f(z) = \ln(z)$ are chosen, one obtains

$$\Lambda = \ln(\underbrace{r[\tfrac{1}{2}\mathbf{p}^2 - (M - \widetilde{m})/r + p_0]}_{\Gamma_0}) - \ln(r[\widetilde{m}/r + m/\Delta]), \tag{6.287}$$

which leads to

$$\frac{dt}{ds} = r/\Gamma_0 \approx r/(rR) = \frac{r\Delta}{\widetilde{m}\Delta + mr}. \tag{6.288}$$

Somewhat surprisingly, the time transformation looks the same. However, in the discrete algorithm, this is not exactly the case and there is a significant difference between the two formulations. While carrying out the Λ_0 step in the first case (6.286), the physical time advances by the amount

$$\Delta t = h_s/\Gamma_0, \quad \Gamma_0 = K + p_0,$$

which means that, to advance the motion, Kepler's equation

$$\Delta t = r_0 X c_1(\beta X^2) + \mathbf{r}_0 \cdot \mathbf{v}_0 X^2 c_2(\beta X^2) + M_{eff} X^3 c_3(\beta X^2) \tag{6.289}$$

must be solved for X, in terms of which the Kepler motion can be advanced by known formulae. Here $\mathbf{r}_0$ and $\mathbf{v}_0$ are the values of coordinates and velocities at the beginning of the step, $\beta = 2M_{eff}/r_0 - v_0^2$ and the c_k are Stumpff functions. In the second case (6.287) it is the quantity $X (= \int_0^{\Delta t} dt/r)$ that advances by the amount $X = h_s/\Gamma_0, \Gamma_0 = r(K + p_0)$, and Kepler's equation gives the corresponding amount of time increment by explicit evaluation of the expression (6.289).

It is to be stressed that the quantities Γ_0 are different in the two formulations by a factor of r. The really significant point is that, in the first case, Kepler's equation must be solved (Δt is given), but in the second case, only the increment of the physical time is computed using Kepler's equation (X is given).

A difficult question here is how to choose the adjustable constant $\widetilde{m}$. Numerical experiments suggest that the best value is $\widetilde{m} = m$, but no obvious reason has been found for this. One should, however, note that if $\widetilde{m} = 0$, and $f(z) = \ln(z)$, then the scaling $dt \propto r$ for small central body distance will not take place. On the other hand, if some other function is used, e.g. $f'(z) = 1/\sqrt{1+(z/m)^2}$, the said scaling remains.

One should note that the two-body motion defined by the Hamiltonian $\Lambda_0 = f(r[\frac{1}{2}\mathbf{p}^2 - (M-\widetilde{m})/r + p_0])$ is independent of the value of $\widetilde{m}$, save for a scaling in the timestep. This occurs because, as explained before, the motion is Keplerian with the effective mass $M_{eff} = r(\frac{1}{2}\mathbf{p}^2 + p_0)$ and this does not depend in any way on the value of $\widetilde{m}$. Thus $\widetilde{m}$ only affects the lengths of the timesteps.

Of course, the leapfrog must also advance the motion due to the remainder R, whichever formulation we use. However, this only consists of the momentum increments

$$\Delta\mathbf{p} = -h_s f'(gR)\frac{\partial(gR)}{\partial\mathbf{r}}, \quad \Delta p_0 = -h_s f'(gR)\frac{\partial(rR)}{\partial t}. \tag{6.290}$$

In this case, however, p_0 will remain constant as R does not contain the time.

Restricted Three-Body Problem

One of the possible forms of the restricted three-body problem Hamiltonian is

$$H = \tfrac{1}{2}\mathbf{p}^2 - M/r - m(1/\Delta - \mathbf{r}\cdot\mathbf{r}_1/r_1^3), \tag{6.291}$$

where $\mathbf{r}_1$ is the position vector of the small mass m with respect to the big mass M. In analogy to the two fixed centres problem, the functional Hamiltonian can be written $\Lambda = \Lambda_0 + \Lambda_1$, where

$$\Lambda_0 = f(r[\tfrac{1}{2}\mathbf{p}^2 - (M-\widetilde{m})/r + p_0]), \tag{6.292}$$

$$\Lambda_1 = -f(r[\widetilde{m}/r + 1/\Delta - \mathbf{r}\cdot\mathbf{r}_1/r_1^3]). \tag{6.293}$$

Again, one could take $f(z) = \ln(z)$ and get results that are very similar to those in the case of the two fixed centres. However, in the problem of two fixed centres (with $f(z) = \ln(z)$), $dt/ds \approx 1/R$, and R is greater than zero. This may not be true in the present case because of the indirect term in the potential, which makes $R < 0$ at large distance. The method breaks down at this point. Another possibility is to choose something else for $f(z)$ or, rather, for its derivative $f'(z)$, in order that $f'(rR)$ remains positive. A reasonable alternative in this case is

$$f'(z) = 1/(1 + z/m + \sqrt{1+(z/m)^2}), \tag{6.294}$$

which reduces, for large z, to $f'(z) \sim m/z$ and the method then behaves, in this limit, in a way similar to the $f(z) = \ln(z)$ alternative. In the case of large negative z, $f \sim 1$, and the time transformation then approaches the simple $dt/ds \approx r$ law. However, at large distances, the physical timestep would depend on the angle between $\mathbf{r}$ and $\mathbf{r}_1$, which does not seem reasonable. Also the physical timestep should not be too large when the distance is large. This is because the perturbation is periodic and the timestep should remain a small fraction of that period.

An alternative is to consider the system in the centre-of-mass frame, because, in this case, the indirect term in the potential does not appear.

Inertial Coordinate System

If the Hamiltonian is written in an inertial coordinate system, such as the centre-of-mass system, then

$$H = \tfrac{1}{2}\mathbf{p}^2 - M/r_1 - m/r_2, \tag{6.295}$$

where r_1 and r_2 are distances measured from M and m, respectively. In this case we may take

$$\Lambda_0 = f(r[\tfrac{1}{2}\mathbf{p}^2 - \widetilde{m}/r + p_0]), \tag{6.296}$$

$$\Lambda_1 = -f(r[M/r_1 + m/r_2 - \widetilde{m}/r]), \tag{6.297}$$

with arbitrary $\widetilde{m}$ (note the different meaning from our earlier usage of the symbol $\widetilde{m}$). Again, Λ_0 always gives the same two-body motion independent of $\widetilde{m}$. An obvious possibility would be $\widetilde{m} = 0$, but then the proximity of the small mass m becomes important in the time transformation only when $r_2 < mr_1/M$, which is too close for small m. The choice $\widetilde{m} \approx M$ may look reasonable, but then similar difficulties occur (perturbing function may become zero and/or negative) as in the M-centric formulation (6.291). In addition, very eccentric orbits are troublesome in this formulation because, near the primary, the perturbations grow large due to the choice of the reference point where there is actually no attracting body at all. Since the centre-of-mass frame formulation presents its own difficulties, it may be concluded that the method is most useful for orbits that remain in the neighbourhood of the primaries.

External Forces

If the use of the proposed method(s) is contemplated in Solar System simulations, a severe problem is encountered: the potential due to the planets, other than the one in the vicinity of which the particle moves, can be so large that the time transformation does not shorten the stepsize early enough and the regularizing effect is largely lost. For example, if a particle is moving near the Earth and Jupiter's perturbation is taken into account, then the Earth and Jupiter potentials become comparable in numerical value at a distance of the order of 1% of the Sun's distance. This is too late and makes the method useless in this case. (Although the Hill sphere of the Earth is only about 0.01 (in our units) the trouble begins in fact at much larger distances.)

To make use of this method, it seems necessary to treat the extra forces in the way suggested for inclusion of non-canonical forces. This means that the method is no longer symplectic, but just composite. It remains, however, time-symmetric and time-reversible, which generally gives good behaviour largely resembling that of symplectic methods. Let **F** be an extra acceleration to be added to the system represented by the Hamiltonian $\Lambda = f(g[K + p_0]) - f(gR)$. The only difference from the symplectic formulation is the extra terms due to **F** in the momentum increment equations. Thus if $t' = gf'(gR)$ then

$$\frac{d\mathbf{p}}{ds} = \frac{\partial f(gR)}{\partial \mathbf{r}} + t'\mathbf{F}, \tag{6.298}$$

$$\frac{dp_0}{ds} = \frac{\partial f(gR)}{\partial t} - \mathbf{p} \cdot (t'\mathbf{F}), \tag{6.299}$$

which must be solved over the s interval h_s to obtain the momentum increments. As the coordinates t and $\mathbf{r}$ are constants in this operation, the (exact) solution is

$$\Delta\mathbf{p} = h_s \left[\frac{\partial f(gR)}{\partial \mathbf{r}} + t'\mathbf{F} \right], \tag{6.300}$$

$$\Delta p_0 = h_s \left[\frac{\partial f(gR)}{\partial t} - \langle \mathbf{p} \rangle \cdot (t'\mathbf{F}) \right], \tag{6.301}$$

where $\langle \mathbf{p} \rangle$ is the average of $\mathbf{p}$ over the step, i.e. $\langle \mathbf{p} \rangle = \mathbf{p}(0) + \Delta\mathbf{p}/2$.

Chapter 7
KS-Regularization

In gravitational problems, regularization means the removal of the $1/r^2$ singularity (when $r \to 0$) using coordinate and time transformations. In this chapter we first discuss the time transformations for regularization methods, followed by the two-body problem regularization using the KS-transformation (Kustaanheimo and Stiefel, 1965). Finally, application of the KS method to multiparticle systems is described. Usually, just the regularized equations of motion are derived, while the Gragg–Bulirsch–Stoer (GBS) extrapolation method for the actual calculation of motions is recommended.

7.1 Extended Phase Space and Time Transformations

As discussed in Section 3.6 a canonical time transformation can be done by using Poincaré's formulation. One adopts

$$\Gamma = g(\mathbf{p}, \mathbf{q})(H(\mathbf{p}, \mathbf{q}, t) + p_0) \tag{7.1}$$

in place of the Hamiltonian. Then the new independent variable s is tied to the physical time t by the canonical equation of motion

$$t' = \frac{\partial \Gamma}{\partial p_0} = g(\mathbf{p}, \mathbf{q}), \tag{7.2}$$

and for p_0 (the momentum of time) the initial value $p_0(0) = -H(0)$ must be taken. The evolution of the system is then governed by the equations

$$\mathbf{p}' = -\frac{\partial \Gamma}{\partial \mathbf{q}}, \quad p_0' = -\frac{\partial \Gamma}{\partial t}, \quad \mathbf{q}' = \frac{\partial \Gamma}{\partial \mathbf{p}}, \quad t' = \frac{\partial \Gamma}{\partial p_0}, \tag{7.3}$$

which guarantee, among other things, that in the correct solution $\Gamma(t) = 0$.

In N-body computations, the time-transformation function $g(\mathbf{p}, \mathbf{q})$ is customarily chosen in such a way that, in the case of a close encounter, $g \sim r$ for very small distance r. Functions having such a quality in the N-body problem are

$$g = 1/U, \quad U = \text{potential}, \tag{7.4}$$

or

$$g = 1/L, \ \ L = \text{Lagrangian}. \tag{7.5}$$

In the case of a conservative system, e.g. an N-body problem, one may write

$$\Gamma = (H - E)/L, \tag{7.6}$$

where the Lagrangian has been chosen for the time transformation and E is the constant, total energy $E = H(0) = -p_0$.

7.2 KS-Transformation

7.2.1 Regularization in One Dimension

Let the Hamiltonian be $H = \frac{1}{2}p^2 - 1/|q|$ and let us carry out the transformation $q = Q^2$. Then the point transformation generating function is

$$S = pq = pQ^2 \tag{7.7}$$

and

$$P = \frac{\partial S}{\partial Q} = 2pQ \tag{7.8}$$

or

$$q = Q^2, \ \ p = \frac{P}{2Q}. \tag{7.9}$$

Thus

$$H = \frac{P^2}{8Q^2} - \frac{1}{Q^2}. \tag{7.10}$$

Poincaré's time transformation $t' = g = q = Q^2$ results in

$$\Gamma = g(H - E) = \tfrac{1}{8}P^2 - 1 - Q^2 E, \tag{7.11}$$

which is a harmonic oscillator Hamiltonian. The singularity was removed by using both a coordinate and a time transformation. The equations of motion

$$Q' = \frac{P}{4}, \quad P' = 2EQ, \quad t' = Q^2 \tag{7.12}$$

have the solution

$$Q = Q_0 \cos(\omega s) + Q'_0 \frac{\sin(\omega s)}{\omega}, \quad (P = 4Q'), \tag{7.13}$$

where $\omega = \sqrt{-E/2}$. If we choose $Q_0 = 0$ (which is valid at some moment), then for the time $t' = (Q'_0/\omega)^2 \sin^2(\omega s)$. This gives

$$t = \left(\frac{Q'_0}{\omega}\right)^2 \left(\frac{1}{2}s - \frac{\sin(2\omega s)}{4\omega}\right), \tag{7.14}$$

which is clearly Kepler's equation and the angle $2\omega s$ is the eccentric anomaly.

7.2.2 Two-Dimensional Regularization

In two dimensions, the Keplerian Hamiltonian is

$$H = \frac{1}{2}\mathbf{p}^2 - \frac{M}{r}, \quad \mathbf{p} = (p_x, p_y), \quad \mathbf{r} = (x, y). \tag{7.15}$$

This case can be regularized with the *Levi-Civita transformation* (Levi-Civita, 1920), which reads, in complex notation,

$$x + iy = (Q_1 + iQ_2)^2 = Q_1^2 - Q_2^2 + i2Q_1Q_2, \tag{7.16}$$

or

$$\begin{aligned} x &= Q_1^2 - Q_2^2, \quad y = 2Q_1Q_2 \quad \text{and} \\ r &= |x + iy| = |Q_1 + iQ_2|^2 = Q_1^2 + Q_2^2. \end{aligned} \tag{7.17}$$

The generating function is

$$S = (Q_1^2 - Q_2^2)p_x + 2Q_1Q_2p_y \tag{7.18}$$

and the new momenta are

$$P_1 = \frac{\partial S}{\partial Q_1} = +2Q_1p_x + 2Q_2p_y, \tag{7.19}$$

$$P_2 = \frac{\partial S}{\partial Q_2} = -2Q_2p_x + 2Q_1p_y, \tag{7.20}$$

or, in matrix form,

$$\mathbf{P} = 2\begin{pmatrix} Q_1 & Q_2 \\ -Q_2 & Q_1 \end{pmatrix} \mathbf{p}. \tag{7.21}$$

The coordinate transformation may also be written in matrix form as

$$\mathbf{r} = \widehat{\mathbf{L}}(\mathbf{Q})\mathbf{Q}, \tag{7.22}$$

where

$$\widehat{\mathbf{L}}(\mathbf{Q}) = \begin{pmatrix} Q_1 & -Q_2 \\ Q_2 & Q_1 \end{pmatrix} \tag{7.23}$$

is the so-called Levi-Civita matrix. With this matrix, the transformation of the momenta can be written

$$\mathbf{P} = 2\widehat{\mathbf{L}}^t(\mathbf{Q})\mathbf{p}. \tag{7.24}$$

Now $\widehat{\mathbf{L}}^t\widehat{\mathbf{L}} = \widehat{\mathbf{L}}\widehat{\mathbf{L}}^t = Q^2\mathbf{I}$, so that multiplication by $\widehat{\mathbf{L}}$ gives

$$\mathbf{p} = \frac{\widehat{\mathbf{L}}(\mathbf{Q})\mathbf{P}}{2Q^2}. \tag{7.25}$$

The square of the momentum $\mathbf{p}$ takes the form

$$\mathbf{p}^2 = \frac{1}{4}\frac{\mathbf{P}^t\widehat{\mathbf{L}}^t(\mathbf{Q})\widehat{\mathbf{L}}(\mathbf{Q})\mathbf{P}}{Q^4} = \frac{1}{4}\frac{P^2}{Q^2}, \tag{7.26}$$

which is the analogue of the one-dimensional case. The new Hamiltonian is similarly

$$\Gamma = Q^2\left(\frac{1}{8}\frac{P^2}{Q^2} - \frac{1}{Q^2} - E\right) = \frac{1}{8}P^2 - 1 - Q^2E. \tag{7.27}$$

Thus the solution is a two-component harmonic oscillator.

7.2.3 Kustaanheimo–Stiefel (KS) Transformation

In three dimensions, a transformation having the same properties as the Levi-Civita transformation does not exist. However, there is a transformation from four-dimensional (Q_1, Q_2, Q_3, Q_4) space to the three-dimensional (x, y, z) space which does have the required properties. This method was published by Kustaanheimo and Stiefel (1965) and it is generally called the KS-transformation. It was based on earlier work by Kustaanheimo (1964) and is also related to the quaternions (Stiefel and Scheifele, 1971; Waldvogel, 2008). Earlier the KS-transformation was used by Hopf (1931) for other purposes. Figure 7.1 contains illustrations of three ways to choose regularized vectors in four-body problems. Similar constructions can be used for more numerous systems, but which one is most suitable is problem-dependent.

In analogy to the Levi-Civita case, one can write

$$\begin{pmatrix} \mathbf{r} \\ 0 \end{pmatrix} = \widehat{\mathbf{Q}}\mathbf{Q}, \tag{7.28}$$

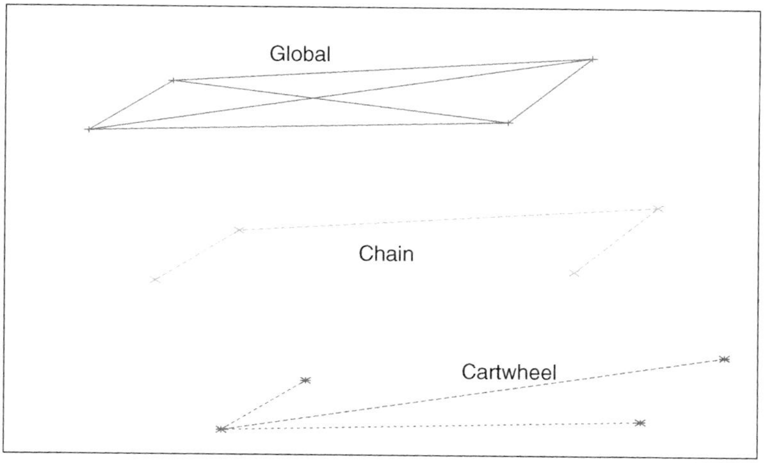

Fig. 7.1. KS regularized interactions in the global method of Heggie, the chain method and Zare's cartwheel method.

where the KS matrix $\widehat{\mathbf{Q}}$ corresponds to the Levi-Civita matrix, and its explicit form is

$$\widehat{\mathbf{Q}} = \begin{pmatrix} Q_1 & -Q_2 & -Q_3 & Q_4 \\ Q_2 & Q_1 & -Q_4 & -Q_3 \\ Q_3 & Q_4 & Q_1 & Q_2 \\ Q_4 & -Q_3 & Q_2 & -Q_1 \end{pmatrix}. \tag{7.29}$$

When we multiply the vector $\mathbf{Q}$ by this matrix the result is

$$\widehat{\mathbf{Q}}\mathbf{Q} = \begin{pmatrix} Q_1Q_1 - Q_2Q_2 - Q_3Q_3 + Q_4Q_4 \\ Q_2Q_1 + Q_1Q_2 - Q_4Q_3 - Q_3Q_4 \\ Q_3Q_1 + Q_4Q_2 + Q_1Q_3 + Q_2Q_4 \\ Q_4Q_1 - Q_3Q_2 + Q_2Q_3 - Q_1Q_4 \end{pmatrix} = \begin{pmatrix} Q_1^2 - Q_2^2 - Q_3^2 + Q_4^2 \\ 2(Q_1Q_2 - Q_3Q_4) \\ 2(Q_1Q_3 + Q_2Q_4) \\ 0 \end{pmatrix}, \tag{7.30}$$

which is a three-dimensional vector because the fourth component is always zero. In scalar form, the KS-transformation may thus be written

$$\begin{aligned} x &= Q_1^2 - Q_2^2 - Q_3^2 + Q_4^2, \\ y &= 2(Q_1Q_2 - Q_3Q_4), \\ z &= 2(Q_1Q_3 + Q_2Q_4). \end{aligned} \tag{7.31}$$

The KS matrix $\widehat{\mathbf{Q}}$ has the same properties as the $\widehat{\mathbf{L}}$ matrix. Among other things,

$$\widehat{\mathbf{Q}}'\widehat{\mathbf{Q}} = \widehat{\mathbf{Q}}\widehat{\mathbf{Q}}' = Q^2\mathbf{I}, \tag{7.32}$$

which gives for the inverse matrix

$$\widehat{\mathbf{Q}}^{-1} = \widehat{\mathbf{Q}}'/Q^2. \tag{7.33}$$

From the relation

$$\mathbf{r} = \widehat{\mathbf{Q}}\mathbf{Q}, \tag{7.34}$$

it follows that $\mathbf{r}^2 = \mathbf{r}^t\mathbf{r} = \mathbf{Q}^t\widehat{\mathbf{Q}}^t\widehat{\mathbf{Q}}\mathbf{Q} = \mathbf{Q}^t(Q^2\mathbf{I})\mathbf{Q} = Q^2\mathbf{Q}^t\mathbf{Q} = Q^4$, or

$$r = Q^2 = Q_1^2 + Q_2^2 + Q_3^2 + Q_4^2. \tag{7.35}$$

In this transformation the number of variables is increased, so that the classical theory of canonical transformations does not apply. One can, however, show that if the point-transformation formulae are (formally) used as before, the result is correct (e.g. Stiefel and Scheifele, 1971). Thus

$$\begin{aligned} S = \mathbf{r}\cdot\mathbf{p} = {} & (Q_1^2 - Q_2^2 - Q_3^2 + Q_4^2)p_x \\ & + 2(Q_1Q_2 - Q_3Q_4)p_y \\ & + 2(Q_1Q_3 + Q_2Q_4)p_z \end{aligned} \tag{7.36}$$

and

$$\begin{aligned} P_1 &= \partial S/\partial Q_1 = +2Q_1p_x + 2Q_2p_y + 2Q_3p_z, \\ P_2 &= \partial S/\partial Q_2 = -2Q_2p_x + 2Q_1p_y + 2Q_4p_z, \\ P_3 &= \partial S/\partial Q_3 = -2Q_3p_x - 2Q_4p_y + 2Q_1p_z, \\ P_4 &= \partial S/\partial Q_4 = +2Q_4p_x - 2Q_3p_y + 2Q_2p_z \end{aligned} \tag{7.37}$$

or

$$\mathbf{P} = 2\begin{pmatrix} Q_1 & Q_2 & Q_3 & Q_4 \\ -Q_2 & Q_1 & Q_4 & -Q_3 \\ -Q_3 & Q_4 & Q_1 & Q_2 \\ Q_4 & -Q_3 & Q_2 & -Q_1 \end{pmatrix}\begin{pmatrix} p_x \\ p_y \\ p_z \\ 0 \end{pmatrix} = 2\widehat{\mathbf{Q}}^t\begin{pmatrix} \mathbf{p} \\ 0 \end{pmatrix}, \tag{7.38}$$

where the last column has been added to make the matrix equal to $\widehat{\mathbf{Q}}^t$ (this does not affect the result because the fourth component of the $\mathbf{p}$ vector is zero in any case).

When the equation

$$\mathbf{P} = 2\widehat{\mathbf{Q}}^t\begin{pmatrix} \mathbf{p} \\ 0 \end{pmatrix}$$

is multiplied by the matrix $\mathbf{Q}$, one obtains

$$\widehat{\mathbf{Q}}\mathbf{P} = 2\widehat{\mathbf{Q}}\widehat{\mathbf{Q}}^t\begin{pmatrix} \mathbf{p} \\ 0 \end{pmatrix} = 2Q^2\begin{pmatrix} \mathbf{p} \\ 0 \end{pmatrix} \tag{7.39}$$

or

$$\begin{pmatrix} \mathbf{p} \\ 0 \end{pmatrix} = \frac{1}{2Q^2}\widehat{\mathbf{Q}}\mathbf{P}. \tag{7.40}$$

Note that this gives for the fourth component

$$P_1Q_4 - P_2Q_3 + P_3Q_2 - P_4Q_1 = 0, \tag{7.41}$$

which is initially valid due to the form of the expression for **P**. (One can show that this relation is also a constant of motion, a fact that is solely the result of the form of the KS-transformation and has nothing to do with the dynamics.) Relation (7.41) is the so-called bilinear relation, and, as it is an integral of motion, its value is always numerically zero. The above expression for **p** gives

$$\mathbf{p}^2 = \frac{\mathbf{P}^t\widehat{\mathbf{Q}}^t\widehat{\mathbf{Q}}\mathbf{P}}{4Q^4} = \frac{P^2}{4Q^2}. \tag{7.42}$$

Thus, if $H = \frac{1}{2}\mathbf{p}^2 - 1/r$, the transformation results in

$$\Gamma = r(H - E) = Q^2\left(\frac{1}{8}\frac{P^2}{Q^2} - \frac{1}{Q^2} - E\right) = \frac{P^2}{8} - 1 - EQ^2, \tag{7.43}$$

which is formally the same as in the one- and two-dimensional cases. The solution for this system gives, via the transformations, the correct solution for the original problem. The additional degrees of freedom do not cause any problems. To find initial values for the components of the **Q** vector, one may select any solution of the equations (7.31). A commonly used possibility is given here. With $\mathbf{R} = (X, Y, Z)^t$, $R = |\mathbf{R}|$, calculate

$$\begin{aligned} u_1 &= \sqrt{\frac{1}{2}(R + |X|)}, \\ u_2 &= Y/(2u_1), \\ u_3 &= Z/(2u_1), \\ u_4 &= 0, \end{aligned} \tag{7.44}$$

then the components of **Q** are

$$\mathbf{Q} = \begin{cases} (u_1, u_2, u_3, u_4)^t, & X \geq 0, \\ (u_2, u_1, u_4, u_3)^t, & X < 0, \end{cases} \tag{7.45}$$

and the values for the momenta are then obtained from (7.38). In Figure 7.2 the difference of the behaviour of the physical and KS coordinates is illustrated for the two-body problem in an almost straight-line orbit. In the figure the thick line represents the physical x coordinate (all others are very near zero) and the thin line with the circles follows the motion of the KS coordinate q_1. The fact that the KS coordinates have no singularities, but at the collision point they move essentially along a straight line, helps in numerical integration. The time transformation which makes the independent

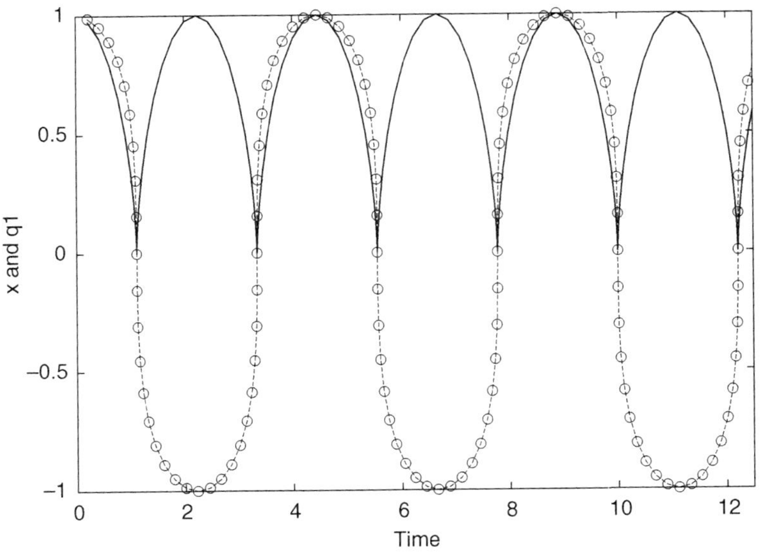

Fig. 7.2. The effect of KS-transformation in the coordinates. The KS coordinates move along a straight line in a point-mass collision orbit.

variable proportional to (the change of) the eccentric anomaly also has an important influence.

7.3 Binaries as Perturbed Harmonic Oscillators

The Hamiltonian for a perturbed binary can be written

$$H = \frac{1}{2}\mathbf{p}^2 - \frac{M}{r} - R(\mathbf{r}, t), \tag{7.46}$$

where M is the total mass of the binary and $R(\mathbf{r}, t)$ is the perturbing function. The KS-transformation with the time-transformation function $g = r = Q^2$ gives the regularized form

$$\Gamma = r(H + p_0) = \tfrac{1}{8}\mathbf{P}^2 + p_0 Q^2 - Q^2 R(\mathbf{r}, t) - M, \tag{7.47}$$

and the equations of motion are

$$\mathbf{P}' = -2p_0\mathbf{Q} + \frac{\partial Q^2 R}{\partial \mathbf{Q}}, \quad p_0' = Q^2 \frac{\partial R}{\partial t}, \quad \mathbf{Q}' = \frac{1}{4}\mathbf{P}, \quad t' = Q^2. \tag{7.48}$$

This form is sometimes useful, especially if the perturbation R is independent of time. However, it is often convenient to introduce the Kepler energy

$$E_K = \tfrac{1}{2}\mathbf{p}^2 - M/r. \tag{7.49}$$

Then $p_0 = -E_K + R$, so that $E'_K = -p'_0 + R' = \mathbf{Q}' \cdot \partial R/\partial \mathbf{Q}$. In addition

$$\frac{\partial R(\mathbf{r}, t)}{\partial \mathbf{Q}} = 2\widehat{\mathbf{Q}}^t \mathbf{F}, \tag{7.50}$$

where $\mathbf{F} = \partial R/\partial \mathbf{r}$ is the (physical) perturbing acceleration. The equations of motion can be reformulated as

$$\mathbf{Q}'' = \frac{E_K}{2}\mathbf{Q} + \frac{1}{2}Q^2\widehat{\mathbf{Q}}^t\mathbf{F}, \tag{7.51}$$

$$E'_K = 2\mathbf{Q}' \cdot \widehat{\mathbf{Q}}^t\mathbf{F}, \tag{7.52}$$

$$t' = Q^2. \tag{7.53}$$

This last form does not explicitly contain the perturbing potential R and it is thus a valid KS-transform for any two-body system with the physical-space equation of motion

$$\ddot{\mathbf{r}} + M\frac{\mathbf{r}}{r^3} = \mathbf{F}, \tag{7.54}$$

irrespective of the origin of the perturbation $\mathbf{F}$ (i.e. the perturbation need not have a potential).

When the perturbation $\mathbf{F}$ is small, the energy E_K changes slowly and equation (7.51) clearly represents a perturbed harmonic oscillator. For such a system, special methods for the computation of the motion can be constructed (e.g. Stiefel and Scheifele, 1971; Mikkola and Aarseth, 1998). For the integration of the time equation $t' = r$, it is important to realize that

$$r'' = 2E_K r + M + r\mathbf{r} \cdot \mathbf{F}, \tag{7.55}$$

which follows easily by the double differentiation of the relation $r = Q^2$ and by use of the energy relation $E_K = (2\mathbf{Q}'^2 - M)/Q^2$. This differential equation for r thus tells us that the derivative of time is also a perturbed harmonic oscillator, but with a double frequency. This fact is used in the next section to construct an approximate solution for the equations of motion.

7.3.1 A Solution Method

The following is a modification of a method presented by Stiefel and Scheifele (1971, pp. 141–144). Let us write the equation of motion for a perturbed harmonic oscillator in the form

$$Y'' + \Omega Y = g(\tau). \tag{7.56}$$

If the function g can be approximated by a polynomial, the Taylor series coefficients can be obtained from

$$Y^{(k+2)} + \Omega Y^{(k)} = g^{(k)}. \tag{7.57}$$

For a polynomial g of (at most) degree n, the recursion reduces to the trigonometric one

$$Y^{(k+2)} + \Omega Y^{(k)} = 0, \tag{7.58}$$

when $k > n$. Thus the expansion for Y may be written

$$\begin{aligned} Y = {} & \sum_{k=0}^{n-2} Y^{(k)} \frac{\tau^k}{k!} \\ & + Y^{(n-1)} \left(\frac{\tau^{n-1}}{(n-1)!} - \frac{\Omega \tau^{n+1}}{(n+1)!} + \frac{\Omega^2 \tau^{n+3}}{(n+3)!} - + \cdots \right) \\ & + Y^{(n)} \left(\frac{\tau^n}{n!} - \frac{\Omega \tau^{n+2}}{(n+2)!} + \frac{\Omega^2 \tau^{n+4}}{(n+4)!} - + \cdots \right), \end{aligned} \tag{7.59}$$

where we recognize the bracketed expressions as the functions $\tau^{n-1} c_{n-1}(\Omega\tau)$ and $\tau^n c_n(\Omega\tau)$.

The simple rule follows. The exact solution for (7.56) can be obtained simply by replacing the inverse factorials in the last two terms by the corresponding c-functions: this gives for Y the expression

$$Y(\tau) = \sum_{k=1}^{n-2} Y^{(k)} \frac{\tau^k}{k!} + Y^{(n-1)} \tau^{n-1} c_{n-1}(\Omega\tau^2) + Y^{(n)} \tau^n c_n(\Omega\tau^2). \tag{7.60}$$

For the oscillator we thus have

$$\mathbf{Q} = \sum_{k=0}^{n-2} \frac{\mathbf{Q}^{(k)} \tau^k}{k!} + \mathbf{Q}^{(n-1)} \tau^{n-1} c_{n-1}(z) + \mathbf{Q}^{(n)} \tau^n c_n(z), \tag{7.61}$$

$$\mathbf{Q}' = \sum_{k=0}^{n-1} \frac{\mathbf{Q}^{(k)} \tau^{k-1}}{(k-1)!} + \mathbf{Q}^{(n-1)} \tau^{n-2} c_{n-2}(z) + \mathbf{Q}^{(n)} \tau^{n-1} c_{n-1}(z), \tag{7.62}$$

$$\delta E_K = \delta E_K' \tau + \frac{\delta E_K''}{2!} \tau^2 + \cdots, \tag{7.63}$$

$$t = \sum_{k=0}^{n-2} \frac{t^{(k)} \tau^k}{k!} + t^{(n-1)} \tau^{n-1} c_{n-1}(4z) + t^{(n)} \tau^n c_n(4z), \tag{7.64}$$

where $z = \Omega_0 \tau^2$ and the derivatives are evaluated at $\tau = 0$. The expression for the time can be deduced from the relation $t' = r$ and the r equation (7.55).

The derivatives in the expressions can usually be determined by repeated differentiation of the equations of motion, or by some Steffensen-like construction (see Section 5.2.3), or by some interpolation-based approximation process for the perturbation part of the oscillator equation. Mikkola and Aarseth (1998) wrote the equation of motion in the form

$$\mathbf{Q}'' + \Omega \mathbf{Q} = \tfrac{1}{2}\delta E_K \mathbf{Q} + \tfrac{1}{2} r \widehat{\mathbf{Q}}^t \mathbf{F}, \tag{7.65}$$

where $\Omega = -\frac{1}{2}E_K(0)$ is a constant, and $\delta E_K = E_K - E_K(0)$ with the derivative $\delta E'_K = E'_K$. They used an interpolation polynomial to approximate the right-hand side of (7.65) and integrated the resulting approximate equation.

As a result of the use of the c-functions, this method is exact for the unperturbed motion. One should also note that, for the two-body energy E_K, the Taylor series approximation is adequate since it is a constant in the unperturbed motion. Computation of the c-functions can be seen in Section 2.6.

7.4 Multiparticle Regularization

7.4.1 Three-Body Problem: AZ Method

Aarseth and Zare (1974) (hereafter AZ) first introduced the idea of applying several KS-transformations simultaneously to a three-body system. This important paper was the beginning of new developments in regularization of multiparticle systems. Since the method is a special case of a later generalization (the chain method), we leave aside most of the theoretical details and just give the main points of the method. The notation adopted here differs from that in the original paper.

Taking the vectors $\mathbf{R}_1$ and $\mathbf{R}_2$ (see Figure 7.3) as new canonical coordinates, the generating function takes the form

$$S = \mathbf{W}_1 \cdot \mathbf{R}_1 + \mathbf{W}_2 \cdot \mathbf{R}_2 = \mathbf{W}_1 \cdot (\mathbf{r}_1 - \mathbf{r}_3) + \mathbf{W}_2 \cdot (\mathbf{r}_2 - \mathbf{r}_3) \tag{7.66}$$

and thus the physical momenta are $\mathbf{p}_1 = \partial S/\partial \mathbf{r}_1 = \mathbf{W}_1$, $\mathbf{p}_2 = \partial S/\partial \mathbf{r}_2 = \mathbf{W}_2$ and $\mathbf{p}_3 = \partial S/\partial \mathbf{r}_3 = -\mathbf{W}_1 - \mathbf{W}_2$.

Application of the KS-transformation to both of the vectors $\mathbf{R}_1$ and $\mathbf{R}_2$ and the momenta $\mathbf{W}_1$ and $\mathbf{W}_2$, accompanied with the time transformation $dt = R_1 R_2 ds = Q_1^2 Q_2^2\, ds$, gives the Hamiltonian in the form

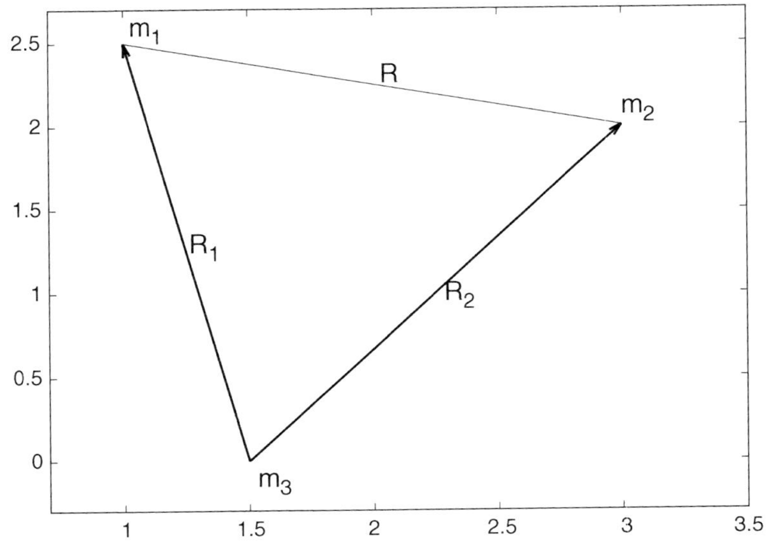

Fig. 7.3. Illustration of the AZ method. The largest distance $R = |\mathbf{R}_2 - \mathbf{R}_1|$ is chosen to be the non-regularized 'singular' distance.

$$\Gamma = \frac{1}{8\mu_{13}} Q_2^2 \mathbf{P}_1^2 + \frac{1}{8\mu_{23}} Q_1^2 \mathbf{P}_2^2 + \frac{1}{4m_3} \mathbf{P}_1^t \widehat{\mathbf{Q}}_1^t \widehat{\mathbf{Q}}_2 \mathbf{P}_2$$
$$- m_1 m_3 Q_2^2 - m_2 m_3 Q_1^2 - m_1 m_2 Q_1^2 Q^2{}_2 / R - E Q_1^2 Q_2^2, \quad (7.67)$$

where $\mu_{k3} = m_k m_3/(m_k + m_3)$, and E is the numerical value of the total energy calculated from the initial values. The equations of motion $\mathbf{P}_k' = -\partial\Gamma/\partial\mathbf{Q}_k$, $\mathbf{Q}_k' = \partial\Gamma/\partial\mathbf{P}_k$ derived from this are regular with respect to collisions between the body m_3 and either one of the two bodies m_1 and m_2 (prime denotes differentiation with respect to the new independent variable s). A switching of the reference body is carried out whenever the singular distance R becomes the smallest. This method has proved itself in numerous simulations and is one of the best methods for treating critical three-body encounters. One of the most famous cases is the so-called Pythagorean problem. This orbit is shown in Figure 7.4. The very first attempt to compute the motion in this system was made long ago by Burrau (1913), but the first attempts were not successful due to irregularities in the equations of motion.

The authors also give additional equations to account for external perturbations. In that case, the three-body energy E is no longer constant, but must be obtained by integration.

7.4.2 Heggie's Global Method

For the sake of simplicity we present here only the version of Heggie's theory which is valid in the centre-of-mass coordinate system.

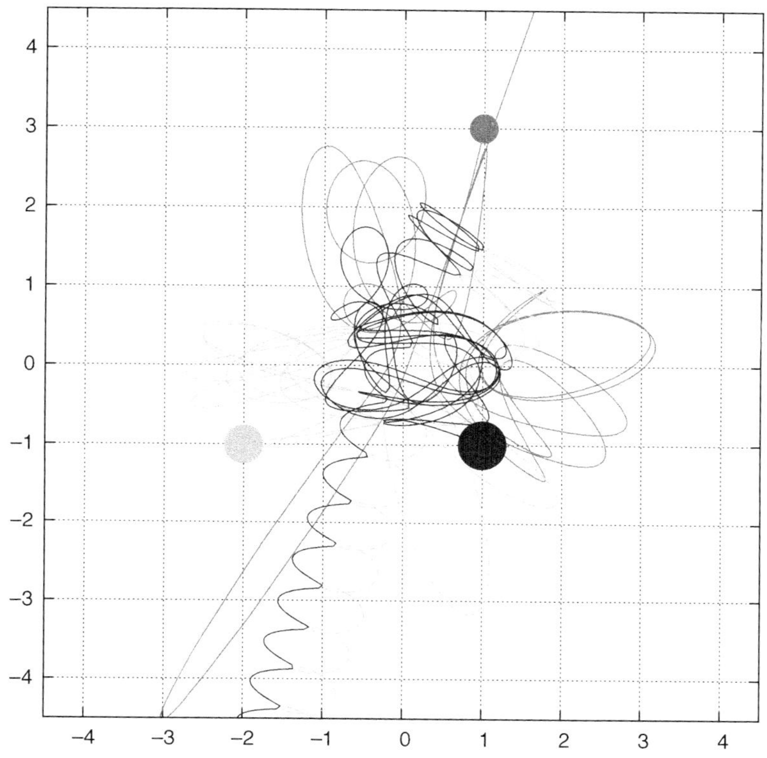

Fig. 7.4. The Pythagorean three-body orbit. This was solved accurately for the first time using the KS-transformation by Szebehely and Peters (1967). The big dots show the starting points. The masses were 3, 4 and 5 units according to the lengths of the opposite triangle sides. The starting velocities were zeros.

Starting with the Hamiltonian

$$H = \sum_{i=1}^{N} \frac{1}{2m_i} w_i^2 - \sum_{i=1}^{N-1} \sum_{j=i+1}^{N} \frac{m_i m_j}{r_{ij}}, \tag{7.68}$$

Heggie (1974) introduces all the relative vectors $\mathbf{r}_{ij} = \mathbf{r}_j - \mathbf{r}_i$ as new dependent variables by using the generating function

$$S = \sum_{i=1}^{N-1} \sum_{j=i+1}^{N} \mathbf{w}_{ij} \cdot (\mathbf{r}_j - \mathbf{r}_i), \tag{7.69}$$

where $\mathbf{w}_{ij}$ are the new momenta conjugate to the new coordinates $\mathbf{r}_{ij}$. A proof for the correctness of such an increase of degrees of freedom is also given in Heggie's (1974) article.

After the introduction of the above transformation, the old momenta are

$$\mathbf{w}_i = \sum_{j=i+1}^{N} \mathbf{w}_{ij} - \sum_{j=1}^{i-1} \mathbf{w}_{ji} \tag{7.70}$$

and the new Hamiltonian in the largely extended phase space takes the form

$$
\begin{aligned}
H(\mathbf{w}_{ij},\mathbf{r}_{ij}) &= \sum_{i=1}^{N-1}\sum_{j=i+1}^{N}\frac{1}{2\mu_{ij}}\mathbf{w}_{ij}^2 \\
&+\sum_{i=1}^{N}\frac{1}{m_i}\left(\sum_{j=i+1}^{N-1}\sum_{k=j+1}^{N}\mathbf{w}_{ij}\cdot\mathbf{w}_{ik}+\sum_{j=1}^{i-2}\sum_{k=j+1}^{i-1}\mathbf{w}_{ji}\cdot\mathbf{w}_{ki}\right. \\
&\left.-\sum_{j=i+1}^{N}\sum_{k=1}^{i-1}\mathbf{w}_{ij}\cdot\mathbf{w}_{ki}\right)-\sum_{i=1}^{N-1}\sum_{j=i+1}^{N}\frac{m_i m_j}{r_{ij}}.
\end{aligned} \tag{7.71}
$$

Application of the KS-transformation to each conjugate pair $(\mathbf{w}_{ij},\ \mathbf{r}_{ij})$ together with the time transformation $dt=\prod_{i=1}^{N-1}\prod_{j=i+1}^{N} r_{ij}ds$ completes the construction of the globally regular Hamiltonian

$$
\begin{aligned}
\Gamma &= [H(\mathbf{w}_{ij},\mathbf{r}_{ij})-E]\prod_{i=1}^{N-1}\prod_{j=i+1}^{N} r_{ij} \\
&= \left[H\left(\frac{\widehat{\mathbf{Q}}_{ij}\mathbf{P}_{ij}}{2Q_{ij}^2},\widehat{\mathbf{Q}}_{ij}\mathbf{Q}_{ij}\right)-E\right]\prod_{i=1}^{N-1}\prod_{j=i+1}^{N} Q_{ij}^2.
\end{aligned} \tag{7.72}
$$

The resulting expressions for the equations of motion are, however, rather complex and the method was not used for more than three bodies until a more concise formulation was provided by Mikkola (1985). That formulation was used for the first time in simulations discussed in Mikkola (1983, 1984a,b).

7.4.3 Algorithm for Global Regularization

Mikkola (1985) published a concise formulation for the computation of the equations of motion for Heggie's global method. The two simple, but essential, steps in the reformulation are as follows:

1. Rename the variables $(\mathbf{w}_{ij},\ \mathbf{r}_{ij})$ to $(\mathbf{p}_k,\ \mathbf{q}_k)$, where $k=(i-1)N-i(i+1)/2+j$. Also write the kinetic energy $T=\sum_{\mu,\nu=1}^{K} T_{\mu\nu}\mathbf{p}_\mu\cdot\mathbf{p}_\nu$ and the potential energy $Q=\sum_{k=1}^{K} M_k/q_k$, where $K=N(N-1)/2$ is the number of $\mathbf{p}$ and $\mathbf{q}$. One easily sees that the constant coefficients $T_{\mu\nu}$ are combinations of inverse masses and the M_k are products of masses. The adopted coordinate system in graphically illustrated in Figure 7.5.
2. Instead of the distance product $\prod_{i<j} r_{ij}$, the inverse of the Lagrangian $1/(T+U)$ is adopted as a time-transformation function. Originally this selection was based on simplicity of formulation, but it has turned out later that the Lagrangian also improves the numerical stability of the method. (This time transformation was in fact suggested by Szebehely and Zare (1975), but not in connection with KS-regularization.)

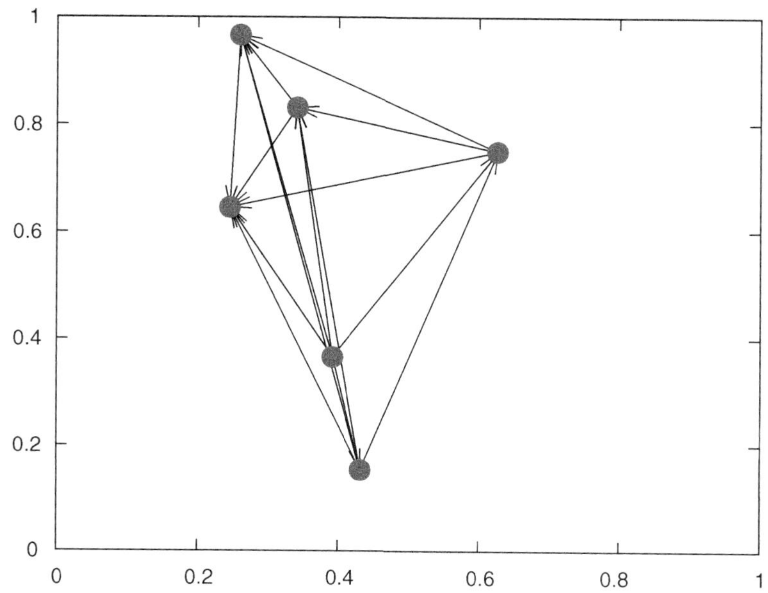

Fig. 7.5. In Heggie's global regularization all the interparticle vectors are treated as new dependent variables.

With the above notation one can write

$$\Gamma = (T - U - E)/(T + U), \tag{7.73}$$

where

$$T = \sum_{\mu\nu} T_{\mu\nu} \mathbf{P}_\mu^t \widehat{\mathbf{Q}}_\mu^t \widehat{\mathbf{Q}}_\nu \mathbf{P}_\nu / (4 Q_\mu^2 Q_\nu^2), \quad U = \sum_k M_k / Q_k^2, \tag{7.74}$$

and the generation of the equations of motion now looks much more tractable than with the double-index formulation. In fact, a simple collection of formulae can be readily given.

(i) Calculate for $1 \leq i < j \leq N$ the initial values and auxiliary quantities

$$\begin{aligned}
k &= (i-1)N - i(i+1)/2 + j, \\
\mathbf{q}_k &= \mathbf{r}_i - \mathbf{r}_j, \\
\mathbf{p}_k &= (m_i \dot{\mathbf{r}}_i - m_j \dot{\mathbf{r}}_j)/N, \\
M_k &= m_i m_j, \\
a_{ik} &= 1, \quad a_{jk} = -1 \ \text{(other components } = 0), \\
T_{\mu\nu} &= \tfrac{1}{2} \textstyle\sum_{n=1}^{N} a_{n\mu} a_{n\nu} / m_n.
\end{aligned} \tag{7.75}$$

Calculate the initial values for the KS variables $\mathbf{P}_k$ and $\mathbf{Q}_k$ according to the usual formulae. After this, the bulk of the numerical effort, the integration of the equations of motion, can be carried out using the formulation below. Here partial derivatives are expressed as subscripts and $\widehat{\mathbf{P}}$ is the analogue of $\widehat{\mathbf{Q}}$.

(ii) Given the KS variables $\mathbf{P}_k$ and $\mathbf{Q}_k$, the derivatives $\mathbf{P}'_k$ and $\mathbf{Q}'_k$ are evaluated as

$$\begin{aligned}
\mathbf{p}_k &= \widehat{\mathbf{Q}}_k \mathbf{P}_k/(2Q_k^2),\\
\mathbf{A}_k &= \textstyle\sum_{\mu=1}^{K} T_{k\mu}\mathbf{p}_\mu,\\
d_k &= A_k \cdot \mathbf{p}_k,\\
T &= \textstyle\sum_{k=1}^{K} d_k, \quad U = \sum_{k=1}^{K} M_k/Q_k^2,\\
T_{\mathbf{P}_k} &= \widehat{\mathbf{Q}}_k^t \mathbf{A}_k/Q_k^2,\\
T_{\mathbf{Q}_k} &= (\widehat{\mathbf{P}}_k^t \mathbf{A}_k^* - 4d_k\mathbf{Q}_k)/Q_k^2,\\
\mathbf{A}_k^* &= (A_1, A_2, A_3, -A_4)_k^t,\\
U_{\mathbf{Q}_k} &= -2M_k\mathbf{Q}_k/Q_k^4,\\
L = T + U, &\quad H = T - U, \quad \Gamma = (H - E)/L,\\
\Gamma_T = (1-\Gamma)/L, &\quad \Gamma_U = -(1+\Gamma)/L.
\end{aligned} \tag{7.76}$$

The derivatives for the KS vectors $\mathbf{P}$ and $\mathbf{Q}$ and the physical time t are now given by

$$\begin{aligned}
\mathbf{Q}'_k &= \Gamma_T T_{\mathbf{P}_k},\\
\mathbf{P}'_k &= -\Gamma_T T_{\mathbf{Q}_k} - \Gamma_U U_{\mathbf{Q}_k},\\
t' &= 1/L.
\end{aligned} \tag{7.77}$$

(iii) Finally the transformation from the KS variables to the physical ones is obtained as follows. For $1 \le i < j \le N,\ k = k(i,j)$ as before, one computes

$$\begin{aligned}
\mathbf{r}_{ij} &= \mathbf{q}_k = \widehat{\mathbf{Q}}_k \mathbf{Q}_k,\\
\mathbf{w}_{ij} &= \mathbf{p}_k = \widehat{\mathbf{Q}}_k \mathbf{P}_k/(2Q_k^2),\\
\mathbf{r}_i &= \left(\textstyle\sum_{j=i+1}^{N} m_j \mathbf{r}_{ij} - \sum_{j=1}^{i-1} m_j \mathbf{r}_{ji}\right) / \sum_{n=1}^{N} m_n,\\
\dot{\mathbf{r}}_i &= \left(\textstyle\sum_{j=i+1}^{N} \mathbf{w}_{ij} - \sum_{j=1}^{i-1} \mathbf{w}_{ji}\right) / m_i.
\end{aligned} \tag{7.78}$$

7.4.4 Chain Regularization

The AZ method was generalized by Mikkola and Aarseth (1990, 1993) for an arbitrary number of bodies. The idea is to form a chain of particles such that all the small interparticle vectors in the system are a part of the chain. Application of the KS-transformation to the chain vectors then gives a regular system of equations of motion.

As the structure of the system continually changes, it is necessary to update the chain, sometimes after every integration step. This method does not increase the degrees of freedom, except in the use of the KS-transformation, and thus it can be used more efficiently for more particles, contrasted with the global regularization.

The N-Body Hamiltonian

Consider a perturbed N-body system with inertial system coordinates $\mathbf{r}_i$, velocities $\mathbf{v}_i$ and masses m_i $(i = 1, \ldots, N)$. We introduce the total mass $M = \sum m_i$, the centre-of-mass coordinates $\mathbf{r}_0 = \sum m_i \mathbf{r}_i / M$ and the centre-of-mass velocity $\mathbf{v}_0 = \sum m_i \mathbf{v}_i / M$. In terms of the coordinates in the centre-of-mass system, $\mathbf{q}_i = \mathbf{r}_i - \mathbf{r}_0$, and the corresponding momenta, $\mathbf{p}_i = m_i(\mathbf{v}_i - \mathbf{v}_0)$, $\mathbf{p}_0 = M\mathbf{v}_0$, the total kinetic energy T_{tot} and the force function U_{tot} are defined by

$$T_{tot} = \frac{1}{2M}\mathbf{p}_0^2 + \sum_{i=1}^{N} \frac{1}{2m_i}\mathbf{p}_i^2, \tag{7.79}$$

$$U_{tot} = \sum_{1 \le i < j \le N} \frac{m_i m_j}{|\mathbf{q}_i - \mathbf{q}_j|} + \widetilde{U}. \tag{7.80}$$

Here $\widetilde{U}$ is a perturbing force function, which may consist of a conservative and a time-dependent part. Note that the centre-of-mass position $\mathbf{r}_0$ is ignorable in the absence of external perturbations, but may appear if external forces are present.

Suppose a chain of vectors connecting N bodies has been selected. After relabelling the bodies such that they are 1, 2, ..., N along the chain, the generating function

$$S = \sum_{i=1}^{N-1} \mathbf{W}_k \cdot (\mathbf{q}_{k+1} - \mathbf{q}_k) \tag{7.81}$$

can be used to obtain the old momenta $\mathbf{p}_k = \partial S / \partial \mathbf{q}_k$ in terms of the new ones. In general, $\mathbf{p}_k = \mathbf{W}_{k-1} - \mathbf{W}_k$, $k = 2, \ldots, N-1$, but the first and last take the form $\mathbf{p}_1 = -\mathbf{W}_1$, $\mathbf{p}_N = \mathbf{W}_{N-1}$. By definition the corresponding chain vectors are given by $\mathbf{R}_k = \mathbf{q}_{k+1} - \mathbf{q}_k$. Omitting the centre-of-mass kinetic energy and the perturbative potential, the corresponding unperturbed Hamiltonian becomes

$$\begin{aligned} H = &\sum_{k=1}^{N-1} \frac{1}{2}\left(\frac{1}{m_k} + \frac{1}{m_{k+1}}\right)\mathbf{W}_k^2 - \sum_{k=2}^{N} \frac{1}{m_k}\mathbf{W}_{k-1} \cdot \mathbf{W}_k \\ &- \sum_{k=1}^{N-1} m_k m_{k+1} / R_k - \sum_{1 \le i \le j-2} m_i m_j / R_{ij}, \end{aligned} \tag{7.82}$$

where the non-chained distances are given by $R_{ij} = |\mathbf{r}_j - \mathbf{r}_i| = |\mathbf{q}_j - \mathbf{q}_i| = |\sum_{i \le k' \le j-1} \mathbf{R}_{k'}|$. Later, different parts of this Hamiltonian, the kinetic energy, the chained and non-chained parts of the force function, respectively, will be denoted by T, U_c and U_{nc}.

Substitution of the KS-transformations $\mathbf{R}_k = \widehat{\mathbf{Q}}_k \mathbf{Q}_k$ and $\mathbf{W}_k = \widehat{\mathbf{Q}}_k \mathbf{P}_k/(2Q_k^2)$ gives the Hamiltonian in terms of the regularizing variables $\mathbf{Q}_k$ and $\mathbf{P}_k$. Here $\widehat{\mathbf{Q}}$ is the KS matrix (see e.g. Stiefel and Scheifele, 1971, p. 24)

$$\widehat{\mathbf{Q}} = \begin{pmatrix} Q_1 & -Q_2 & -Q_3 & Q_4 \\ Q_2 & Q_1 & -Q_4 & -Q_3 \\ Q_3 & Q_4 & Q_1 & Q_2 \\ Q_4 & -Q_3 & Q_2 & -Q_1 \end{pmatrix}. \tag{7.83}$$

In most applications it is sufficient to retain only the three upper rows of the matrix (cf. AZ). Here, however, the 4×4 matrices are used throughout and it is to be understood in the standard way of adding a fourth zero component to the physical vector. The abbreviation $\widehat{\mathbf{P}}$ will be used for the analogue of $\widehat{\mathbf{Q}}$, in which the components of $\mathbf{Q}$ are replaced by components of $\mathbf{P}$. The corresponding transposed quantities are denoted by $\widehat{\mathbf{Q}}^t$ or $\widehat{\mathbf{P}}^t$. After applying the time transformation $dt = g\,ds$, with $g = 1/(T + U)$, one obtains the regularized Hamiltonian $\Gamma = g(H - E) = (T - U - E)/(T + U)$ in the ($\mathbf{P}$, $\mathbf{Q}$, s) system (cf. Mikkola and Aarseth, 1993). Here E is the (numerical) value of the Hamiltonian H evaluated from the initial conditions. The derivation of the equations of motion $\mathbf{P}_k{}' = -\partial\Gamma/\partial\mathbf{Q}_k$, $\mathbf{Q}_k{}' = \partial\Gamma/\partial\mathbf{P}_k$, where primes denote differentiation with respect to s, is now possible remembering that $\partial\widetilde{U}/\partial\mathbf{Q} = 2\widehat{\mathbf{Q}}^t\mathbf{F}$, where $\mathbf{F} = \partial\widetilde{U}/\partial\mathbf{R}$.

Chain Selection and Updating

The process is started by finding the shortest interparticle vector. This is taken as the first part of the chain under construction. Next, one searches for the particle which is closest to one or the other end of the presently known part of the chain. Once identified, this particle is added to the chain (to the end to which it is closer). This operation is repeated until all particles are included in the chain. In practice this can be done efficiently by first sorting the distances and searching sequentially the still unused distances (one can also skip those distances that would form a loop within the known part of the chain).

After every integration step the chain is checked for need of updating. First, one checks whether any non-chained part is the shortest in all those triangles formed by two consecutive chain

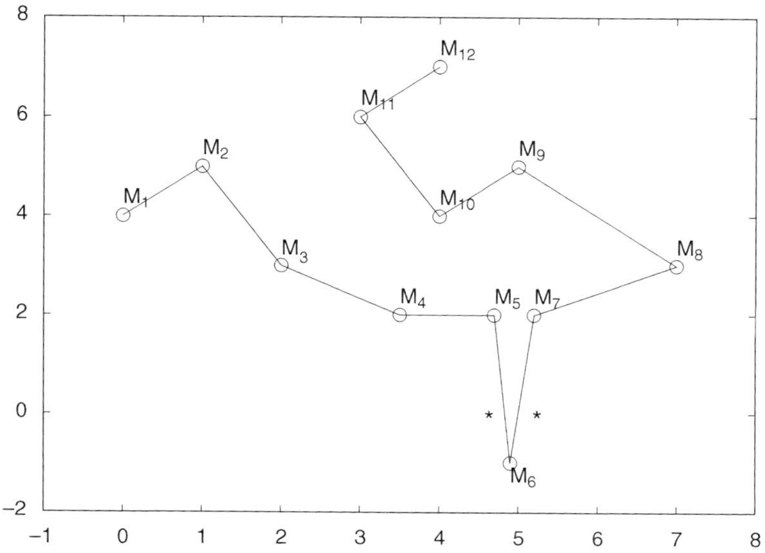

Fig. 7.6. Illustration of the chain and the checking of switching conditions. Distances like $R_{5,7}$ are compared with the smaller of the two distances $R_{5,6}$ and $R_{6,7}$ (marked by *).

vectors and one non-chained vector, in which case the chain is switched. Switching is also necessary if any other non-chained vector is shorter than the smallest of the chained vectors that are in contact with one or the other end of the vector under consideration. However, to avoid unnecessary switching operations, checking those distances that are larger than (say) twice the average length of a chain vector can be ignored.

Figure 7.6 illustrates the meaning of the conditions stated above in the case of a 12-particle chain. To avoid some potential roundoff problems (due to large distance ratios) particularly in the case of close two- or multi-body configurations, it is advantageous to carry out the transformation from the old chain vectors $\mathbf{R}_k$ to the new ones directly by expressing the new chain vectors as sums of the old ones.

Let the actual 'physical' names of the chain particles 1, ..., N (as defined above) be I_1, I_2, ..., I_N and let us use the notation I_k^{old} and I_k^{new} for the names in the old and new chains. Then,

$$\mathbf{q}_{I_k^{old}} = \sum_{\nu=1}^{k-1} \mathbf{R}_\nu^{old}, \tag{7.84}$$

$$\mathbf{R}_\mu^{new} = \mathbf{q}_{I_{\mu+1}^{new}} - \mathbf{q}_{I_\mu^{new}}. \tag{7.85}$$

Thus one needs to use the correspondence between the old and new indices to express the new chain vector $\mathbf{R}$ in terms of the old ones. One easily finds that, if k_0 and k_1 are two indices such that $I_{k_0}^{old} = I_\mu^{new}$ and $I_{k_1}^{old} = I_{\mu+1}^{new}$, then

$$\mathbf{R}_{\mu}^{new} = \sum_{\nu=1}^{N-1} B_{\mu\nu}\mathbf{R}_{\nu}^{old}, \tag{7.86}$$

where $B_{\mu\nu} = +1$ if ($k_1 > \nu$ and $k_0 \leq \nu$) and $B_{\mu\nu} = -1$ if ($k_1 \leq \nu$ and $k_0 > \nu$), otherwise $B_{\mu\nu} = 0$.

External Perturbations

Let $\mathbf{F}_j$ be the perturbing acceleration acting on body m_j in the inertial coordinate system. The system may be written in Hamiltonian form by adding the perturbing potential

$$\delta U = \sum_{j=1}^{N} m_j \mathbf{q}_j \cdot (\mathbf{F}_j(t) - \mathbf{F}_0(t)) + M\mathbf{r}_0 \cdot \mathbf{F}_0(t), \tag{7.87}$$

where $\mathbf{F}_0 = \sum_{j=1}^{N}(m_j/M)\mathbf{F}_j$ is the centre-of-mass acceleration in the inertial system.

From the transformation formulae, one easily derives the recursive equations

$$\delta\dot{\mathbf{p}}_j = m_j(\mathbf{F}_j - \mathbf{F}_0), \;\; j = 1, \ldots, N, \tag{7.88}$$

$$\delta\dot{\mathbf{W}}_1 = -\delta\dot{\mathbf{p}}_1, \tag{7.89}$$

$$\delta\dot{\mathbf{W}}_k = \delta\dot{\mathbf{W}}_{k-1} - \delta\dot{\mathbf{p}}_k, \;\; k = 2, \ldots, N-2, \tag{7.90}$$

$$\delta\dot{\mathbf{W}}_{N-1} = \delta\dot{\mathbf{p}}_N, \tag{7.91}$$

where δ denotes the part of the derivative which is due to the perturbation. The corresponding corrections to the derivatives of the KS momenta are then given by

$$\delta\mathbf{P}_k{}' = 2g\widehat{\mathbf{Q}}_k^t \delta\dot{\mathbf{W}}_k, \tag{7.92}$$

and the rate of change of the internal energy may be written

$$E' = 2\sum_{k=1}^{N-1} \mathbf{Q}_k'^t \widehat{\mathbf{Q}}_k^t \delta\dot{\mathbf{W}}_k. \tag{7.93}$$

Algorithms

In this section the formulae necessary for transforming from physical variables to chain and chain KS variables as well as the inverse procedure are provided. The analogous, but modified, transformations used when switching the chain are also given. Finally, the expressions for the derivatives are summarized. The formulation presented here is based on the time transformation $g = 1/L$, where $L = T + U$ is the Lagrangian.

Transformations

In this subsection the transformation formulae are given. Contrary to our notation above, 'true' indices for the physical coordinates and velocity vectors are used, instead of those assigned while relabelling the bodies after finding the appropriate chain.

(i) Transformation to regularized variables. First, calculate all the distances in the system and determine the chain, indicating the location of body m_j by an index I_j such that the actual index (name) of the jth chain particle is I_j. The centre of mass and chain variables are then calculated according to

$$\mathbf{r}_0 = \sum_{j=1}^{N} \frac{m_j}{M}\mathbf{r}_j, \quad \dot{\mathbf{r}}_0 = \sum_{j=1}^{N} \frac{m_j}{M}\dot{\mathbf{r}}_j, \tag{7.94}$$

$$\mathbf{p}_j = m_{I_j}(\dot{\mathbf{r}}_{I_j} - \dot{\mathbf{r}}_0), \; j = 1, \ldots, N, \tag{7.95}$$

$$\mathbf{W}_1 = -\mathbf{p}_1, \tag{7.96}$$

$$\mathbf{W}_k = \mathbf{W}_{k-1} - \mathbf{p}_k, \; k = 2, \ldots, N-2, \tag{7.97}$$

$$\mathbf{W}_{N-1} = \mathbf{p}_N, \tag{7.98}$$

$$\mathbf{R}_k = \mathbf{r}_{I_{k+1}} - \mathbf{r}_{I_k}, \; k = 1, \ldots, N-1. \tag{7.99}$$

KS-transformations for $\widehat{\mathbf{Q}}_k\mathbf{Q}_k = \mathbf{R}_k$ are obtained in the standard way. Defining
$\mathbf{R}_k = (X, Y, Z)^t$ and $R = |\mathbf{R}_k|$, one may then use the formulae

$$\begin{aligned} u_1 &= \sqrt{\frac{1}{2}(R + |X|)}, \\ u_2 &= Y/(2u_1), \\ u_3 &= Z/(2u_1), \\ u_4 &= 0 \end{aligned} \tag{7.100}$$

and the components of $\mathbf{Q}_k$ are

$$\mathbf{Q}_k = \begin{cases} (u_1, u_2, u_3, u_4)^t, & X \geq 0, \\ (u_2, u_1, u_4, u_3)^t, & X < 0. \end{cases} \tag{7.101}$$

The related KS momenta are given by

$$\mathbf{P}_k = 2\widehat{\mathbf{Q}}_k^t \mathbf{W}_k. \tag{7.102}$$

(ii) Transformation to physical variables. The transformation from $\mathbf{Q}, \mathbf{P}$ to $\mathbf{r}, \dot{\mathbf{r}}$ begins by the inverse KS mapping of the chain vectors and the corresponding momenta:

$$\mathbf{R}_k = \widehat{\mathbf{Q}}_k \mathbf{Q}_k, \tag{7.103}$$

$$\mathbf{W}_k = \widehat{\mathbf{Q}}_k \mathbf{P}_k/(2Q_k^2), \quad k = 1, \ldots, N-1, \tag{7.104}$$

$$\mathbf{p}_1 = -\mathbf{W}_1, \tag{7.105}$$

$$\mathbf{p}_k = \mathbf{W}_{k-1} - \mathbf{W}_k; \quad k = 2, \ldots, N-1, \tag{7.106}$$

$$\mathbf{p}_N = \mathbf{W}_{N-1}, \tag{7.107}$$

$$\dot{\mathbf{r}}_{I_j} = \mathbf{p}_j/m_j + \dot{\mathbf{r}}_0, \quad j = 1, \ldots, N. \tag{7.108}$$

To obtain the coordinates $\mathbf{r}_j$, one first forms coordinates $\mathbf{q}_j$ in a system where $\mathbf{q}_1 = 0$, which are then converted to the physical coordinates. This gives rise to the expressions

$$\mathbf{q}_1 = 0, \quad \mathbf{q}_{j+1} = \mathbf{q}_j + \mathbf{R}_j, \quad j = 1, \ldots, N-1, \tag{7.109}$$

$$\mathbf{q}_0 = \sum_{j=1}^{N} \frac{m_j}{M} \mathbf{q}_j, \tag{7.110}$$

$$\mathbf{r}_{I_j} = \mathbf{q}_j - \mathbf{q}_0 + \mathbf{r}_0, \quad j = 1, \ldots, N. \tag{7.111}$$

(iii) Switching. For each value of $\mu = 1, \ldots, N-1$, find the indices k_0 and k_1 such that $I_{k_0}^{old} = I_{\mu}^{new}$ and $I_{k_1}^{old} = I_{\mu+1}^{new}$. With the help of these indices, write

$$\mathbf{R}_{\mu}^{new} = \sum_{\nu=1}^{N-1} B_{\mu\nu} \mathbf{R}_{\nu}^{old}, \tag{7.112}$$

where $B_{\mu\nu} = +1$ if ($k_1 > \nu$ and $k_0 \le \nu$) and $B_{\mu\nu} = -1$ if ($k_1 \le \nu$ and $k_0 > \nu$), otherwise $B_{\mu\nu} = 0$. To transform the chain momenta, first obtain the physical momenta from the old chain by using

$$\mathbf{p}_{I_1^{old}} = -\mathbf{W}_1, \tag{7.113}$$

$$\mathbf{p}_{I_k^{old}} = \mathbf{W}_{k-1} - \mathbf{W}_k, \quad k = 2, \ldots, N-1, \tag{7.114}$$

$$\mathbf{p}_{I_N^{old}} = \mathbf{W}_{N-1}, \tag{7.115}$$

and evaluate the new chain momenta by using

$$\mathbf{W}_1 = -\mathbf{p}_{I_1^{new}}, \tag{7.116}$$

$$\mathbf{W}_k = \mathbf{W}_{k-1} - \mathbf{p}_{I_k^{new}}, \quad k = 2, \ldots, N-2, \tag{7.117}$$

$$\mathbf{W}_{N-1} = \mathbf{p}_{I_N^{new}}. \tag{7.118}$$

Note that here the same notation was used for new and old chain momenta.

Evaluation of Derivatives

Define the matrix T_{ij} and the mass products M_k and M_{ij} by

$$T_{kk} = \tfrac{1}{2}(1/m_k + 1/m_{k+1}), \tag{7.119}$$

$$T_{k\,k+1} = -1/m_k, \quad k = 1, \ldots, N-1, \tag{7.120}$$

$$M_k = m_k m_{k+1}, \tag{7.121}$$

$$M_{ij} = m_i m_j, \quad i < j. \tag{7.122}$$

Further auxiliary variables are introduced by

$$\mathbf{W}_k = \widehat{\mathbf{Q}}_k \mathbf{P}_k/(2Q_k^2), \tag{7.123}$$

$$\mathbf{A}_k = \sum \tfrac{1}{2}(T_{ki} + T_{ik})\mathbf{W}_i, \quad |i-k| \le 1, \quad 1 \le i \le N-1, \tag{7.124}$$

$$D_k = \mathbf{A}_k \cdot \mathbf{W}_k, \tag{7.125}$$

$$T = \sum D_k, \quad 1 \le k \le N-1, \tag{7.126}$$

$$\mathbf{R}_{ij} = \mathbf{q}_j - \mathbf{q}_i = \sum \mathbf{R}_{k'}, \quad i \le k' \le j-1, \tag{7.127}$$

$$U_{nc} = \sum M_{ij}/R_{ij}, \quad 1 \le i \le j-2, \tag{7.128}$$

$$U = \sum_{k=1}^{N-1} M_k/R_k + U_{nc}, \tag{7.129}$$

$$g = 1/(T+U). \tag{7.130}$$

When calculating the vectors $\mathbf{R}_{ij}$ above, the difference form is generally more efficient, but the summation form should be used if there are only two terms in the sum; this avoids some roundoff problems. Writing partial derivatives (gradients) with subscripts,

$$T_{\mathbf{P}_k} = \widehat{\mathbf{Q}}_k^t \mathbf{A}_k/Q_k^2, \tag{7.131}$$

$$\mathbf{A}_k^* = (A_1, A_2, A_3, -A_4)^t, \tag{7.132}$$

$$T_{\mathbf{Q}_k} = (\widehat{\mathbf{P}}_k^t \mathbf{A}_k^* - 4D_k \mathbf{Q}_k)/Q_k^2, \tag{7.133}$$

$$U_{nc\mathbf{R}_k} = -\sum M_{ij}\mathbf{R}_{ij}/R_{ij}^3, \quad i \le k \le j-1, \tag{7.134}$$

$$U_{nc\mathbf{Q}_k} = 2\widehat{\mathbf{Q}}_k^t U_{nc\mathbf{R}_k}, \tag{7.135}$$

$$U_{\mathbf{Q}_k} = -2M_k\mathbf{Q}_k/Q_k^4 + U_{nc\mathbf{Q}_k}. \tag{7.136}$$

If perturbations are present, the physical positions (and velocities, if needed) are obtained according to the transformation formulae given above, whereupon the external acceleration $\mathbf{F}_j$ acting on each particle can be calculated. For the centre-of-mass acceleration,

$$\mathbf{F}_0 = \sum_{j=1}^{N} \frac{m_j}{M} \mathbf{F}_j. \tag{7.137}$$

With the perturbative contributions

$$\delta \dot{\mathbf{p}}_j = m_j(\mathbf{F}_j - \mathbf{F}_0), \;\; j = 1, \ldots, N, \tag{7.138}$$

$$\delta \dot{\mathbf{W}}_1 = -\delta \dot{\mathbf{p}}_1, \tag{7.139}$$

$$\delta \dot{\mathbf{W}}_k = \delta \dot{\mathbf{W}}_{k-1} - \delta \dot{\mathbf{p}}_k, \;\; k = 1, \ldots, N-2, \tag{7.140}$$

$$\delta \dot{\mathbf{W}}_{N-1} = \delta \mathbf{p}_N, \tag{7.141}$$

$$\delta \mathbf{P}_k{}' = 2g\widehat{\mathbf{Q}}_k^t \delta \dot{\mathbf{W}}_k, \tag{7.142}$$

the equations of motion for KS variables can be written

$$H = T - U, \;\; \Gamma = g(H - E), \tag{7.143}$$

$$\Gamma_T = (1 - \Gamma)g, \;\; \Gamma_U = -(1 + \Gamma)g, \tag{7.144}$$

$$\Gamma_{\mathbf{P}_k} = \Gamma_T T_{\mathbf{P}_k}, \tag{7.145}$$

$$\Gamma_{\mathbf{Q}_k} = \Gamma_T T_{\mathbf{Q}_k} + \Gamma_U U_{\mathbf{Q}_k}, \tag{7.146}$$

$$\mathbf{P}_k{}' = -\Gamma_{\mathbf{Q}_k} + \delta \mathbf{P}_k{}', \tag{7.147}$$

$$\mathbf{Q}_k{}' = \Gamma_{\mathbf{P}_k}. \tag{7.148}$$

The centre-of-mass coordinates and velocities may be advanced by

$$\mathbf{v}_0' = g\mathbf{F}_0, \tag{7.149}$$

$$\mathbf{r}_0' = g\mathbf{v}_0, \tag{7.150}$$

while the rates of change for the internal energy and time are

$$E' = 2 \sum_{k=1}^{N-1} \mathbf{Q}_k'^t \widehat{\mathbf{Q}}_k^t \delta \dot{\mathbf{W}}_k, \tag{7.151}$$

$$t' = g. \tag{7.152}$$

7.4.5 External Potential and Additional Forces

If the external forces arise from a time-independent conservative potential $U_{ext} = U_{ext}(\mathbf{r}_1, \ldots, \mathbf{r}_{SSN})$, then it is not necessary to integrate the value of the internal energy E, but one may use the total energy conservation to obtain it. If U_{ext} gives rise to a centre-of-mass acceleration, then the change of the centre-of-mass kinetic energy is to be taken into account so that

$$E = E(0) + T_{cm} - U_{ext} - (T_{cm}(0) - U_{ext}(0)). \tag{7.153}$$

Here $E(0)$ is the initial value of the internal energy, calculated only from the Newtonian N-body potential and the kinetic energy in the centre-of-mass reference frame. The kinetic energy of the centre of mass is $T_{cm} = \frac{1}{2}Mv_0^2$, and the argument(s) (0) in the above equation denote the initial values.

If there are additional non-conservative forces in the system, then they must simply be added to the forces arising from any conservative potential. However, now the energy change due to this part of the external force must also be integrated. One simply uses the above expression for the rate of energy change, but only the additional non-conservative part of the force is used in this integration. This additional energy variation should then be added to the expression for the internal energy. A set of important external forces are the relativistic post-Newtonian ones. These are given in Section 9.1.

We note that it is possible to treat the conservative external force in just the same way as the non-conservative one. Thus, in many cases, it may be easier to ignore the concept of external potential. However, if the only external force is conservative or the additional non-conservative part is very small in comparison, then precision is usually gained by using for E in the equations of motion the total energy, instead of only the internal energy of the N-body system (Stiefel and Scheifele, 1971).

7.4.6 Stepsize Control and Time Transformations

In numerical integration of the regularized equations of motion, the Bulirsch and Stoer (1966) extrapolation method is useful. This, like a number of other methods, provides estimates of its local error. An integration step is deemed to be successful if this error is less than a predefined tolerance. It is customary to use a relative error check ($|\delta Y/Y| \leq \epsilon$ for every dependent variable Y) for stepsize control. This is usually satisfactory, but there are situations in which it is not possible, and not necessary, to use such a relative accuracy. This may occur, for example, if the system starts at rest. In such a case the integration often stops by shrinking the stepsize to zero.

To avoid tolerance problems, an energy error check can be used instead. The error in the value of the regularized Hamiltonian is

$$\begin{aligned}\delta\Gamma &= \sum\left(\frac{\partial\Gamma}{\partial\mathbf{P}_k}\cdot\delta\mathbf{P}_k + \frac{\partial\Gamma}{\partial\mathbf{Q}_k}\cdot\delta\mathbf{Q}_k\right)\\ &= \sum(\mathbf{Q}_k'\cdot\delta\mathbf{P}_k - \mathbf{P}_k'\cdot\delta\mathbf{Q}_k).\end{aligned} \tag{7.154}$$

Thus it is possible to use the energy conservation check without additional computational overhead, because the coefficients in the energy error estimate are readily available. In practice, it is required that every individual term in the above expression is, in absolute value, less than the specified tolerance. (Note that the meaning of the tolerance is not the same as in a relative precision check, i.e. it must be scaled appropriately.)

Assuming that the coordinates and momenta have certain relative errors, it is easy to see that the energy error may, in the worst case, be comparable to the product of the Lagrangian and the relative errors of the variables. Thus it is reasonable, and possible, to require that the error in the physical Hamiltonian satisfies

$$|\delta H/L| < \epsilon, \tag{7.155}$$

where $L = T + U$ is the Lagrangian. This is actually the condition imposed when the Lagrangian time transformation $g = 1/L$ is used and the error in Γ is checked as discussed above. If other time transformations are used, then it is necessary to modify the conditions to the form

$$|q'\delta p| < \epsilon\, t'L, \quad |p'\delta q| < \epsilon\, t'L, \tag{7.156}$$

where q and p are used as generic abbreviations for coordinates and momenta. In these conditions, the Lagrangian may be replaced by any function of similar magnitude.

An interesting experimental alternative to the time transformation $g = 1/L$ is given by

$$g = 1/(\alpha T + \beta U), \tag{7.157}$$

where α and β are two constants. These may be conveniently normalized by taking $\alpha + \beta = 2$ to make the sum the same as in the case of $g = 1/L$. The only changes in the algorithms are in the expressions

$$g = 1/(\alpha T + \beta U), \quad \Gamma_T = (1 - \alpha\Gamma)g, \quad \Gamma_U = -(1 + \beta\Gamma)g, \tag{7.158}$$

replacing those given above.

7.4.7 N-Body Kepler's Equation

The time-transformed N-body problem, which we have been considering here, has one additional known integral. This can be obtained by considering the quantity

$$S = \sum qp,$$

where q and p represent the physical coordinates and momenta, respectively. Differentiating S gives

$$S' = \sum(q'p + p'q) = g\sum(T_p p + U_q q) = g(2T - U), \quad (7.159)$$

where the last equality follows from the fact that T is a harmonic function of degree two with respect to the momenta, while U is a harmonic function of degree minus one with respect to the coordinates. Applying the relations $E = T - U$, for the time-transformation function $g = 1/(\alpha T + \beta U)$ (where α and β are constants), and eliminating T and U, there results

$$S' = \frac{1}{\alpha+\beta} + \left(1 + \frac{\beta}{\alpha+\beta}\right) Eg. \quad (7.160)$$

As $g = t'$ integrating the above equation gives

$$\Delta \sum qp = \frac{1}{\alpha+\beta}\Delta s + \left(1 + \frac{\beta}{\alpha+\beta}\right) E\Delta t, \quad (7.161)$$

where the Δ signify the increments of the quantities following them. This equation thus provides a relation between the scalar product of momenta and coordinates, the independent variable s and the physical time t. The existence of this relation is a consequence of the increase of degrees of freedom when the time is effectively a coordinate and the independent variable is a dummy. The integral does not, however, give any essential new information but may be used for accuracy checking purposes. In fact, if the relation for the case of a two-body problem is considered taking for the constants the values $\alpha = 0$, $\beta = 1$, then the expression is nothing but the Kepler's equation. Thus for the class of time transformations $g = 1/(\alpha T + \beta U)$ there exists a kind of N-body Kepler's equation. This may also be used to obtain the time directly without integrating equation (7.152).

Obtaining the corresponding result for the KS-transformed systems is simple: for any physical vector pair $(\mathbf{q}, \mathbf{p})$ and the corresponding KS pair $(\mathbf{Q}, \mathbf{P})$,

$$\mathbf{p} \cdot \mathbf{q} = \frac{\mathbf{P}^t\widehat{\mathbf{Q}}^t\widehat{\mathbf{Q}}\mathbf{Q}}{2Q^2} = \frac{1}{2}\mathbf{P} \cdot \mathbf{Q}. \quad (7.162)$$

Thus the relation $\sum qp = \frac{1}{2}\sum QP$ applies between the physical and KS variables regardless of how many vectors are in question. For example, the relation (7.160) takes the form

$$\Delta \sum PQ = \Delta s + 3E\Delta t \quad (7.163)$$

for the KS-regularized system with $\alpha = \beta = 1$.

7.5 Slow-Down

Mikkola and Aarseth (1996) suggested a method to speed up the computation of the motions of weakly perturbed binaries in N-body systems. The method is not restricted to N-body computations, but may be used whenever motions of binaries under the influence of weak and slowly changing perturbations are studied.

7.5.1 General Principle

Let

$$\ddot{\mathbf{r}} = -M\mathbf{r}/r^3 + \mathbf{F} \tag{7.164}$$

be the equation for the relative motion of the components of a binary. Here $\mathbf{F}$ is a small perturbing acceleration. To speed up the computation, one could multiply $\mathbf{F}$ by a factor κ and say that one period corresponds actually to κ periods. This way the secular perturbations are obtained correctly, but short-periodic perturbations are magnified. Alternatively, the motion of the binary may be slowed down:

$$\begin{aligned} \dot{\mathbf{v}} &= -\kappa^{-1} M\mathbf{r}/r^3 + \mathbf{F}, \\ \dot{\mathbf{r}} &= \kappa^{-1}\mathbf{v}. \end{aligned} \tag{7.165}$$

Here κ is the slow-down coefficient. In this new system, the slowed-down period is now κ-fold compared to the original one, so that the numerical integration is speeded up accordingly.

In the earlier Hamiltonian formulation

$$H = \tfrac{1}{2}p^2 - M/r - U, \tag{7.166}$$

where U is the perturbing force function. The modified formulation may be described by the new Hamiltonian

$$H = \kappa^{-1}(\tfrac{1}{2}p^2 - M/r) - U, \tag{7.167}$$

which gives the above slowed-down equations of motion.

More generally, one may write

$$H_{new} = \kappa^{-1} H_B + (H - H_B), \tag{7.168}$$

where H_B is the Hamiltonian for the relative motion of a binary in an N-body system with total Hamiltonian H.

Considering κ to be a function of time, one may write $\dot{E} = (\partial\kappa^{-1}/\partial t)\, H_B$. When κ is changed stepwise, its time derivative is a delta-function and $\dot{E} = \delta(t - t_c)(1/\kappa^{new} - 1/\kappa^{old})H_B$, where t_c is the moment of change of κ. This can be immediately integrated to get the change of energy

$$\delta E = (1/\kappa^{new} - 1/\kappa^{old})\,(H_B)_{|t=t_c}. \tag{7.169}$$

7.5.2 Application to Chain Regularization

In terms of the chain variables $\mathbf{R}_k$ and $\mathbf{W}_k$, the Hamiltonian can be written in the concise form

$$H = \sum T_{ij}\mathbf{W}_i \cdot \mathbf{W}_j - \sum M_k/R_k - \sum M_{ij}/R_{ij}, \tag{7.170}$$

where the auxiliary quantities affected by the slow-down procedure are

$$\begin{aligned} T_{kk} &= \tfrac{1}{2}(1/m_k + 1/m_{k+1}), \\ T_{k\;k+1} &= -1/m_k, \quad k = 1, \ldots, N-1, \\ M_k &= m_k m_{k+1}. \end{aligned} \tag{7.171}$$

To construct the expression for H_B within the chain, we first note that the relevant term in the potential energy is M_b/R_b if b is the index of the distance between the components $(m_b,\, m_{b+1})$ of the binary. It is clear that we must select from the kinetic energy those terms having $1/m_b$ or $1/m_{b+1}$ as a factor. The resulting combination of terms, however, also contains the centre-of-mass kinetic energy of the binary. Therefore, this latter contribution must be subtracted and added to the rest of the Hamiltonian.

The centre-of-mass kinetic energy for a pair $(m_b,\, m_{b+1})$ is given by

$$T_{cm} = \tfrac{1}{2}(\mathbf{W}_{b+1} - \mathbf{W}_{b-1})^2/(m_b + m_{b+1}).$$

Selecting the relevant terms from the total Hamiltonian, and expanding the above expression for T_{cm}, gives the following new formulation for the evaluation of the matrix T_{ij} and the vector M_k in the presence of a slowed-down binary. Instead of the previous expressions, we now have

$$\begin{aligned} T_{k\;k+1} &= -1/m_k/\kappa_k, \quad k = 1, 3, 5, \ldots, \leq N, \\ T_{k+1\;\;k+2} &= -1/m_{k+1}/\kappa_k, \\ T_{b-1\;\;b+1} &= -(1-\kappa_b^{-1})/(m_b + m_{b+1}), \\ \delta T_{b-1\;\;b-1} &= \delta T_{b+1\;\;b+1} = -\tfrac{1}{2} T_{b-1\;\;b+1}, \\ T_{kk} &= \tfrac{1}{2}(-T_{k\;\;k+1} - T_{k+1\;\;k+2}) + \delta T_{kk}, \\ M_k &= m_k m_{k+1}/\kappa_k, \\ \delta E &= -\frac{m_b m_{b+1}}{2a}(1/\kappa_b^{new} - 1/\kappa_b^{old}), \end{aligned} \tag{7.172}$$

where a slow-down coefficient κ_k has been defined for every chain vector, but taken to be $= 1$, except for $k = b$, and a denotes the semi-major axis of the binary.

After expressing the Hamiltonian in terms of the modified T_{ij} and M_k and the application of the KS-transformations, one gets the equations of motion in the usual way from the new Hamiltonian Γ_{new} by

$$\dot{\mathbf{P}} = -\partial\Gamma_{new}/\partial\mathbf{Q}, \dot{\mathbf{Q}} = \partial\Gamma_{new}/\partial\mathbf{P}. \tag{7.173}$$

It is obvious that the algorithm is almost the same as for the original chain method. The only difference is that the auxiliary constants have modified values and there are a few more non-zero components in the matrix T_{ij}.

For decision-making purposes an (over)estimate of the relative perturbation due to tidal forces (at the maximum apocentre $2a$) can be obtained by using

$$\gamma = 8\frac{a^3}{m_B}\textstyle\sum_j m_j/r_{bj}^3. \tag{7.174}$$

Here the summation extends over all other bodies except the components of the binary. The distances are computed from one or the other of the binary components, a practical process, satisfactory for decision-making. The expression

$$\kappa = \max(1, (\gamma_0/\gamma)^{1/2}) \tag{7.175}$$

can be used for κ. For the theoretical motivation of this, see Mikkola and Aarseth (1996).

Chapter 8
Algorithmic Regularization

8.1 The Basic Leapfrog Algorithms

Because the simple integration method, the leapfrog, is of utmost importance in algorithmic regularization, this chapter discusses it first. The leapfrog is also generally an important method to solve dynamical equations of motion, if not directly, then together with some extrapolation methods (Gragg, 1965; Bulirsch and Stoer, 1966). In these the leapfrog is used first for the substeps, then the methods attempt to extrapolate the results to the zero-stepsize outcome. Often this procedure gives also high accuracy up to the specified error tolerance (often something like 10^{-10} or even 10^{-12}).

Let the differential equations be

$$\ddot{\mathbf{r}} = \mathbf{F}(\mathbf{r}). \tag{8.1}$$

Denoting the velocity of a mass point ($\dot{\mathbf{r}}$) by $\mathbf{v}$, then the motion, over a short timestep h, can be approximated by first moving the point $\mathbf{r}$ linearly over a half-step $h/2$, then by changing the velocity according to the acceleration $\mathbf{F}$, and advancing the position again as

$$\mathbf{r}_{1/2} = \mathbf{r}_0 + \mathbf{v}_0 h/2, \tag{8.2}$$

$$\mathbf{v}_1 = \mathbf{v}_0 + h\mathbf{F}(\mathbf{r}_{1/2}), \tag{8.3}$$

$$\mathbf{r}_1 = \mathbf{r}_{1/2} + \mathbf{v}_1 h/2. \tag{8.4}$$

This way the coordinates and velocities at time $t = t_0 + h$ have been computed. Repeating the sequence of operations, the coordinates and velocities at later times are obtained in a stepwise process.

More generally, let $\mathbf{r}_0 = \mathbf{r}(0)$, $\mathbf{v}_0 = \mathbf{v}(0)$ be the initial values at $t = 0$. Then the algorithm to approximate the motion over a large number of steps up to the time $t_n = nh$ becomes

$$\mathbf{r}_{1/2} = \mathbf{r}_0 + \mathbf{v}_0 h/2, \tag{8.5}$$

$$\mathbf{v}_1 = \mathbf{v}_0 + h\mathbf{F}(\mathbf{r}_{1/2}), \tag{8.6}$$

$$\mathbf{r}_{3/2} = \mathbf{r}_{1/2} + \mathbf{v}_1 h, \tag{8.7}$$

$$\mathbf{v}_2 = \mathbf{v}_1 + h\mathbf{F}(\mathbf{r}_{3/2}), \tag{8.8}$$

$$\vdots \tag{8.9}$$

$$\mathbf{r}_{n-1/2} = \mathbf{r}_{n-3/2} + h\mathbf{v}_{n-1}, \tag{8.10}$$

$$\mathbf{v}_n = \mathbf{v}_{n-1} + h\mathbf{F}(\mathbf{r}_{n-1/2}), \tag{8.11}$$

$$\mathbf{r}_n = \mathbf{r}_{n-1/2} + \mathbf{v}_n h/2. \tag{8.12}$$

Thus the half-steps appear only in the beginning and at the end of the computation because the linear motion at the end of previous elementary leapfrog steps can be concatenated with the starting half-step of the next sequence of operations. The method is of the second order in a long integration (one-step error $O(h^3)$), i.e. if integrated over a long interval of time with shorter and shorter steps, the error decreases as $O(h^2)$. The process is also time-reversible (if round-off error can be neglected), which makes it possible to apply the GBS (Gragg–Bulirsch–Stoer) extrapolation method (Chapter 5 and Section 5.3).

8.1.1 Leapfrog with Adaptive Timestep

Normally leapfrog is used with a constant timestep, because this often gives good long-term behaviour of the numerical solution. To a great extent this positive outcome is due to the fact that leapfrog is a time-reversible integrator (Funato et al., 1996). However, if the orbit approaches a singularity, then a shorter timestep is necessary. If the change of stepsize is done based on a criterion that leads to loss of time-reversibility, secular energy errors usually occur. Huang and Leimkuhler (1997) described a time-reversible way to reduce the timestep according to a given scaling function $g(\mathbf{r}, \mathbf{v})$. The principle may be written

$$\rho_{n+1} + \rho_n = \frac{2}{g(\mathbf{r}_{n+1/2}, \mathbf{v}_{n+1/2})}, \tag{8.13}$$

where ρ is a timestep scaling factor. One notes that the above relation is time-reversible and that the scaling factor ρ follows approximately the evolution of the function g, provided that the starting value is chosen such that strong fluctuations are avoided.

The implementation of this may be done in several ways, for example,

$$\mathbf{r}_{1/2} = \mathbf{r}_0 + \frac{h}{2\rho_0}\mathbf{v}_0, \tag{8.14}$$

$$\mathbf{v}_{1/2} = \mathbf{v}_0 + \frac{h}{2\rho_0}\mathbf{F}_{1/2}, \tag{8.15}$$

$$\rho_1 = \frac{2}{g(\mathbf{r}_{1/2}, \mathbf{v}_{1/2})} - \rho_0, \tag{8.16}$$

$$\mathbf{v}_1 = \mathbf{v}_{1/2} + \frac{h}{2\rho_1}\mathbf{F}_{1/2}, \tag{8.17}$$

$$\mathbf{r}_1 = \mathbf{r}_{1/2} + \frac{h}{2\rho_1}\mathbf{v}_1. \tag{8.18}$$

If the scaling function $g = g(\mathbf{r})$, i.e. depends only on the position, then the two velocity updates can be combined to get $\mathbf{v}_1 = \mathbf{v}_0 + h(1/\rho_0 + 1/\rho_1)/2\mathbf{F}_{1/2}$. There are also other modifications that may be used: one may update ρ in the beginning of a step, instead of in the middle, which changes the role of the two half-steps in the above formulation. For the update of the scaling factor there are also other options, such as $1/\rho_{n+1} + 1/\rho_n = 2g_{n+1/2}$ or, in principle, one may try any symmetrical combination of the two ρ in place of the above suggested ones.

Numerical experiments with the Kepler orbit, and with $\rho = r$, show that this method is a significant improvement over the standard fixed-stepsize leapfrog, especially for highly eccentric orbits.

8.2 Generalized Composite Leapfrog

Consider a differential equation in which the derivative can be conveniently divided into two parts,

$$\dot{\mathbf{y}} = \mathbf{A}(\mathbf{y}) + \mathbf{B}(\mathbf{y}), \tag{8.19}$$

such that the differential equations

$$\dot{\mathbf{y}} = \mathbf{A}(\mathbf{y}), \tag{8.20}$$

$$\dot{\mathbf{y}} = \mathbf{B}(\mathbf{y}) \tag{8.21}$$

can be easily solved (preferably analytically), but the solution of the original equation (8.19) is difficult.

The idea is to concatenate the A and B steps, i.e. start by moving the system using the A-part equations only (over a timestep), and then use the result as initial values for the following B-part motion. More specifically, if $\mathbf{y} = \mathbf{Y}_A(\mathbf{y}_0, t)$ is the solution of the equation $\dot{\mathbf{y}} = \mathbf{A}(\mathbf{y})$ with the initial value $\mathbf{y}_0$, and $\mathbf{Y}_B(\mathbf{y}_0, t)$ is the solution of $\dot{\mathbf{y}} = \mathbf{B}(\mathbf{y})$, then an approximation for the solution of $\dot{\mathbf{y}} = \mathbf{A}(\mathbf{y})+\mathbf{B}(\mathbf{y})$, at $t = h$, is produced in two steps as

$$\mathbf{y}_A = \mathbf{Y}_A(\mathbf{y}_0, h), \quad \mathbf{y}(h) \approx \mathbf{Y}_B(\mathbf{y}_A, h). \tag{8.22}$$

Repeating this procedure leads to an approximate solution of the problem over a longer time. One notices that this simplest *composite method* resembles the so-called Euler method for the solution of differential equations.

An important special case is $\ddot{\mathbf{r}} = \mathbf{F}(\mathbf{r})$. Here one may take for $\mathbf{y}$ the vector $\mathbf{y} = (\mathbf{v}, \mathbf{r})$, where $\mathbf{v} = \dot{\mathbf{r}}$ and the A and B parts can be defined as $\mathbf{A}(\mathbf{y}) = (\mathbf{F}(\mathbf{r}), \mathbf{0})$, $\mathbf{B}(\mathbf{y}) = (\mathbf{0}, \mathbf{v})$. Thus the step of using the A part means solving

$$\dot{\mathbf{v}} = \mathbf{F}(\mathbf{r}), \quad \dot{\mathbf{r}} = \mathbf{0},$$

which gives

$$\mathbf{v} = \mathbf{v}(0) + t\mathbf{F}(\mathbf{r}_0), \quad \mathbf{r} = \text{constant},$$

and in the B part,

$$\dot{\mathbf{r}} = \mathbf{v}, \quad \dot{\mathbf{v}} = 0,$$

thus giving

$$\mathbf{r} = \mathbf{r}_0 + t\mathbf{v}_0, \quad \mathbf{v} = \text{constant}.$$

A convenient notation for the formal solution of differential equations of the above form can be obtained by considering the time derivative of an (arbitrary) function $f(\mathbf{y}(t))$ when $\mathbf{y}(t)$ is the solution for $\dot{\mathbf{y}} = \mathbf{C}(\mathbf{y})$. We have $\dot{f} = \dot{\mathbf{y}} \cdot \partial f(\mathbf{y})/\partial \mathbf{y} = \mathbf{C} \cdot \partial f(\mathbf{y})/\partial \mathbf{y}$ and thus in general

$$f^{(n)} = \left(\mathbf{C} \cdot \frac{\partial}{\partial \mathbf{y}}\right)^n f(\mathbf{y}), \tag{8.23}$$

and the Taylor expansion may be written

$$f(\mathbf{y}(t)) = \sum_{n=0}^{\infty} \frac{f^{(n)}}{n!} t^n = e^{t\mathbf{C}\cdot\partial/\partial\mathbf{y}} f(\mathbf{y})_{|\mathbf{y}=\mathbf{y}_0}. \tag{8.24}$$

The solution of the various parts may thus be written in the operator form

$$\text{A}: \quad \mathbf{y}(h) = e^{h\mathbf{A}(\mathbf{y})\cdot\partial/\partial\mathbf{y}}\mathbf{y}(0), \tag{8.25}$$

$$\text{B}: \quad \mathbf{y}(h) = e^{h\mathbf{B}(\mathbf{y})\cdot\partial/\partial\mathbf{y}}\mathbf{y}(0). \tag{8.26}$$

Using combinations of these A and B parts with various stepsizes it is possible to obtain improved precision, much like in the Runge–Kutta methods. In general, one may symbolize a composite method by the operator

$$e^{X_1 h\widehat{A}} e^{W_1 h\widehat{B}} e^{X_2 h\widehat{A}} \ldots e^{W_n h\widehat{B}} e^{X_{n+1} h\widehat{A}}, \tag{8.27}$$

where, in general, $\widehat{C} = \mathbf{C}\cdot\partial/\partial\mathbf{y}$ and thus $e^{\tau\widehat{C}}$ implies: *move the system forward over a timestep of length* τ *using the differential equation* $\dot{\mathbf{y}} = \mathbf{C}(\mathbf{y})$.

The simplest such method is

$$e^{h\widehat{A}} e^{h\widehat{B}}, \tag{8.28}$$

which is a first-order integrator. If this operator is applied over a half-step $h/2$ and concatenated with a similar operation in which the order of $\widehat{A}$ and $\widehat{B}$ is reversed, a time-symmetric method results:

$$e^{(1/2)h\widehat{A}} e^{h\widehat{B}} e^{(1/2)\widehat{A}}. \tag{8.29}$$

This is a second-order integrator (one-step error $= O(h^3)$), the *generalized leapfrog*. There is also the other leapfrog

$$e^{(1/2)h\widehat{B}} e^{h\widehat{A}} e^{(1/2)\widehat{B}}, \tag{8.30}$$

where the order of the A and B operations has been reversed.

8.3 Algorithmic Regularization

Mikkola and Tanikawa (1999a,b, 2013b) and also Preto and Tremaine (1999) discussed a modification of the leapfrog algorithm which, when applied to two-body equations, gives the exact trajectory and only has an error in time. The properties of this method allow generalization (to the N-body problem), which is discussed in this section.

(In this chapter the letter B is used for the binding energy (i.e. $B = -E$). B is also the 'momentum of time' in the Hamiltonian formulation.)

8.3.1 The Logarithmic Hamiltonian

Let $\mathbf{p}$ be the momenta of the coordinates $\mathbf{q}$. Let further $T(\mathbf{p})$ be the kinetic energy and $U(\mathbf{q}, t)$ the force function such that the Hamiltonian is $H = T - U$. If the time t is also considered to be a coordinate and the corresponding momentum is B, then for this system the function

$$\Lambda = \log(T + B) - \log(U) \tag{8.31}$$

can be used as a Hamiltonian in the extended phase space. The equations of motion derivable from Λ are

$$\mathbf{p}' = -\Lambda_{\mathbf{q}} = U_{\mathbf{q}}/U, \tag{8.32}$$

$$B' = -\Lambda_t = U_t/U, \tag{8.33}$$

$$\mathbf{q}' = \Lambda_{\mathbf{p}} = T_{\mathbf{p}}/(T + B), \tag{8.34}$$

$$t' = \Lambda_B = 1/(T + B), \tag{8.35}$$

where the prime denotes differentiation with respect to the (new) independent variable (call it s) and partial derivatives are denoted by subscripts. As Λ does not depend explicitly on the independent variable s, the value of Λ is constant. Thus we have $B = -T + U +$ constant. If the constant in this expression is chosen to be 0, then $T + B = U$. Using this and dividing the above equations of motion by the equation for time (8.35), the time derivatives become

$$\dot{\mathbf{p}} = U_{\mathbf{q}}, \quad \dot{B} = U_t \quad \text{and} \quad \dot{\mathbf{q}} = T_{\mathbf{p}},$$

which is the standard form of canonical equations for the Hamiltonian $H = T - U$. Thus, using Λ as the Hamiltonian is equivalent to using H, assuming that one takes initially $B(0) = -H(0)$.

The logarithmic Hamiltonian leapfrog algorithm solves the two-body problem with only a phase error. This is true independent of orbital eccentricity and thus it provides an *algorithmic* regularization. This applies also in problems in which there are many $1/r$-type singularities, such as the few-body problem. A new surprising result is that a simple modification of the method produces exact trajectories also for the quasi-Keplerian problem with an $1/r^2$ term in the potential.

8.4 Logarithmic Hamiltonian Derivations

8.4.1 Extended Phase Space and Time Transformations

Poincaré's time-transformed extended phase space Hamiltonian can be written

$$\Gamma = g(p, q)(H(p, q, t) + B),$$

where B is the momentum of time (also binding energy), for which the initial value $B(0) = -H(0)$ guarantees that $\Gamma = 0$ along the orbit. The equations of motion are

$$\begin{aligned} p' &= -\frac{\partial \Gamma}{\partial q} = -g\frac{\partial H}{\partial q} - \frac{\partial g}{\partial q}(H+B), \\ B' &= -g\frac{\partial H}{\partial t}, \\ q' &= \frac{\partial \Gamma}{\partial p} = +g\frac{\partial H}{\partial p} + \frac{\partial g}{\partial p}(H+B), \\ t' &= \frac{\partial \Gamma}{\partial B} = g, \end{aligned} \tag{8.36}$$

where $H + B = 0$ along the orbit and thus the obtained equations are clearly correct because the derivatives are simply multiplied by the time-transformation function g. One must stress that the terms containing the factor $H + B$ cannot be dropped although they are zeros along the exact solution (if dropped, the resulting equations would not be canonical any more). Also, numerical methods are always discrete and thus the relation $H + B = 0$ is not strictly valid during the computational operations and thus the results would be different if the terms were dropped.

Mikkola–Tanikawa Derivation

One can often write the Hamiltonian in the separable form

$$H = T(\mathbf{p}) - U(\mathbf{r}, t), \tag{8.37}$$

with the time-transformation function

$$g = 1/U. \tag{8.38}$$

One obtains the new Hamiltonian Γ as

$$\Gamma = (T - U + B)/U. \tag{8.39}$$

Since $\Gamma = 0$ along the correct orbit, one can use the Hamiltonian

$$\Lambda = \log(1 + \Gamma) \quad (= \Gamma + O(\Gamma^2)), \tag{8.40}$$

which can be simplified to

$$\Lambda = \log(T + B) - \log(U). \tag{8.41}$$

This is the logarithmic Hamiltonian as derived by Mikkola and Tanikawa (1999a,b).

Due to the separability it is possible to use the leapfrog algorithm for approximating the motion.

Preto–Tremaine Derivation
The logarithmic Hamiltonian can also be derived by

$$g = (f(T+B) - f(U))/(T+B-U), \tag{8.42}$$

where f is any function. This looks like a $0/0$ form, but when $B \to -H$, the limit is a well-defined value. Thus one obtains the new Hamiltonian $\widetilde{\Gamma}$ as

$$\widetilde{\Gamma} = g(T+B-U) = f(T+B) - f(U). \tag{8.43}$$

This was proposed by Preto and Tremaine (1999).

With $f(z) = \log(z)$ one arrives again at the logarithmic Hamiltonian.

Derivation by Time Transformation
Using the energy conservation $T + B = U$ one can write

$$ds = (T+B)dt = Udt, \tag{8.44}$$

which gives the equations

$$t' = 1/(T+B), \quad \mathbf{r}' = \frac{\partial T}{\partial \mathbf{p}}/(T+B), \tag{8.45}$$

$$B' = \frac{\partial U}{\partial t}/U, \quad \mathbf{p}' = \frac{\partial U}{\partial \mathbf{r}}/U. \tag{8.46}$$

The first equation ($ds = (T+B)dt$) can be used to transform the time and coordinate equations and the second one ($ds = Udt$) to transform the momentum and energy equations. The results are the canonical equations of the logarithmic Hamiltonian

$$\Lambda = \log(T+B) - \log(U). \tag{8.47}$$

8.5 Logarithmic Leapfrog

In this section the basic form of the algorithm is considered. After that, application to the two-body motion is discussed and finally the generalized leapfrog (in the sense of Wisdom and Holman (1991)) is scrutinized.

The equations for the basic leapfrog are

$$\begin{aligned}
&\delta t = \tfrac{1}{2}h/(T+B), \quad t \to t + \delta t, \quad \mathbf{r} \to \mathbf{r} + \delta t\, \partial T/\partial \mathbf{p},\\
&\delta \tilde{t} = h/U, \quad B \to B + \delta \tilde{t}\, \partial U/\partial t, \quad \mathbf{p} \to \mathbf{p} + \delta \tilde{t}\, \partial U/\partial \mathbf{r},\\
&\delta t = \tfrac{1}{2}h/(T+B), \quad t \to t + \delta t, \quad \mathbf{r} \to \mathbf{r} + \delta t\, \partial T/\partial \mathbf{p}.
\end{aligned} \tag{8.48}$$

In this algorithm the physical timestep becomes small when the potential U ($= T + B$) becomes large. This also gives more:

the results are regular in the case of point-mass collisions in the gravitational few-body problem. Because of the time-symmetrical nature of the leapfrog algorithm the extrapolation methods (Gragg, 1965; Bulirsch and Stoer, 1966) can be used for high accuracy.

8.5.1 Two-Body Leapfrog

To introduce the method, consider the simple case of two-body motion. The logarithmic Hamiltonian takes the form

$$\Lambda = \log(\mathbf{p}^2/2 + B) + \log(r), \tag{8.49}$$

where a constant term (log of mass) has been discarded. The value of B remains constant and must be initially obtained as

$$B = -(\mathbf{p}^2/2 - \mu/r),$$

where μ is the mass of the two-body system. The time transformation is

$$dt = ds\,\frac{\partial \Lambda}{\partial B} = \frac{ds}{(\mathbf{p}^2/2 + B)}. \tag{8.50}$$

Here s is the new independent variable, which is

$$s = \int^t (\mathbf{p}^2/2 + B)\,dt = \int^t \frac{\mu}{r}\,dt, \tag{8.51}$$

i.e. a quantity proportional to the eccentric anomaly. With stepsize h and initial values $\mathbf{p}_0$, $\mathbf{r}_0$, t_0, the leapfrog algorithm takes the form

$$\mathbf{r}_{1/2} = \mathbf{r}_0 + \frac{h}{2}\mathbf{p}_0 \bigg/ \left(\frac{\mathbf{p}_0^2}{2} + B\right), \tag{8.52}$$

$$\mathbf{p}_1 = \mathbf{p}_0 - h\,\mathbf{r}_{1/2}/r_{1/2}^2, \tag{8.53}$$

$$\mathbf{r}_1 = \mathbf{r}_{1/2} + \frac{h}{2}\mathbf{p}_1 \bigg/ \left(\frac{\mathbf{p}_1^2}{2} + B\right), \tag{8.54}$$

$$t_1 = t_0 + \frac{h}{2}\left[1 \bigg/ \left(\frac{\mathbf{p}_0^2}{2} + B\right) + 1 \bigg/ \left(\frac{\mathbf{p}_1^2}{2} + B\right)\right], \tag{8.55}$$

where $\mathbf{p}_1$, $\mathbf{r}_1$, t_1 are the values at the end of the step. It can be demonstrated (Mikkola and Tanikawa, 1999a; Preto and Tremaine, 1999) that *this algorithm actually produces correct positions and momenta on the associated Keplerian ellipse*. However, the values of time are not correct and the method thus has phase errors. An illustration of the working of the method is given in Figure 8.1.

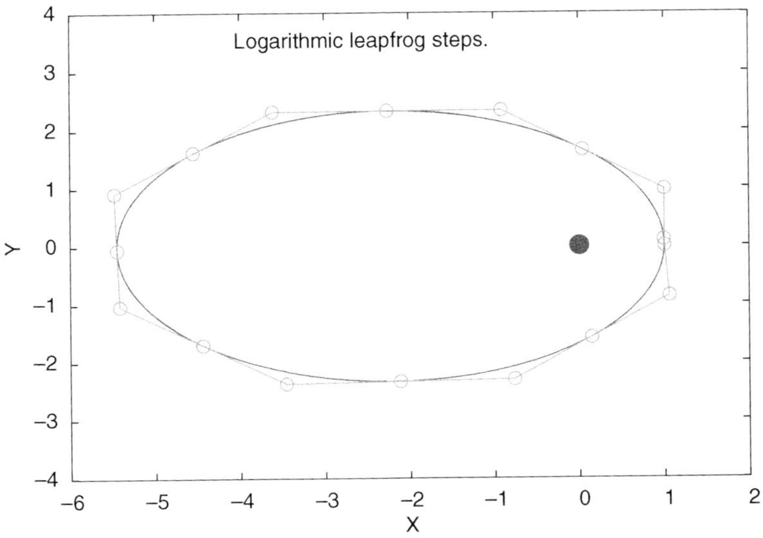

Fig. 8.1. The logarithmic leapfrog gives correct results after every step. In this orbit the eccentricity = 0.69. The midpoint values (outside the orbit) are mathematical quantities and do not give physical positions.

In any case, all the orbital elements, except pericentre time, are exactly conserved by this modified leapfrog algorithm. This is true even for collision orbits where the eccentricity $e = 1$, despite the fact that the momentum $\mathbf{p}$ has a singularity as $|\mathbf{p}| \to \infty$. However, as long as the algorithm does not evaluate the above expressions at $r_{1/2} = 0$, the coordinates stay on the correct trajectory and the values of $\mathbf{p}$ are also the correct ones, corresponding to the computed position.

One sees that in this algorithm the physical timestep becomes short when the potential U $(= T + B)$ becomes large. As known (see below) this also gives more: the results are regular in the case of point-mass collisions in the gravitational few-body problem. Due to the time-symmetrical nature of the leapfrog algorithm, the extrapolation methods (Gragg, 1965; Bulirsch and Stoer, 1966) can then be used for high accuracy.

One strange-looking feature in the above equations is that the mass does not appear explicitly in the equations. It is, however, hidden in the value of the binding energy B. If the mass is not explicitly known, one may obtain it as $m = r(\frac{1}{2}\mathbf{p}^2 + B)$ when needed.

Time Correction in Two-Body Motion

Since the algorithm gives the exact trajectory it can be seen as another way of writing the solution of the two-body problem. To really do that write-up, we need to know the correction for the time value obtained from the algorithm. The correction

$$t_{correct} \to t + \tau(h, m, B), \tag{8.56}$$

as well as the real meaning of the stepsize h in the leapfrog formulae, were derived by Preto and Tremaine (1999) (also in Mikkola and Wiegert, 2002) by comparing the leapfrog expressions with the analytical two-body formulae. One notes that the time correction depends only on the stepsize h, mass m and binding energy B. Thus the correction is the same for every step when these quantities are kept constant. In fact, with $Z = \sqrt{B}h/(\sqrt{2}m)$ we have $\tau = -h^3(Z - \mathrm{atan}(Z))/(4m^2Z^3)$. After some modification of notation the results can be written in terms of h_s, the real stepsize of the independent variable in the differential equations (8.45), as

$$\zeta = h_s^2 B/(2m^2), \tag{8.57}$$

$$h = \frac{m}{\sqrt{B/2}}\tan(\sqrt{\zeta}) = h_s c_1(\zeta)/c_0(\zeta), \tag{8.58}$$

$$\begin{aligned}\tau &= -\frac{h_s^3}{4m^2}\frac{\tan(\sqrt{\zeta}) - \sqrt{\zeta}}{\sqrt{\zeta}^3}\\ &= -\frac{h_s^3}{4m^2}(c_2(\zeta) - c_3(\zeta))/c_0(\zeta).\end{aligned} \tag{8.59}$$

Here c_0, c_1, c_2 and c_3 are the Stumpff c-functions (Stumpff, 1959)

$$c_n(z) = \sum_{k=0}^{\infty}\frac{z^k}{(n+2k)!} \quad \left(= \frac{1}{n!} - zc_{n+2}(z)\right), \tag{8.60}$$

the use of which made it possible to write the formulae in universal form, valid for every orbit type, i.e. including parabolic and hyperbolic orbits. One notes that the time correction depends only on the stepsize, energy and mass. Thus it is the same for every step as long as the stepsize is kept constant. For the computation of the c_n-functions there are instructions, for example, in Mikkola and Aarseth (1998) and in Section 2.6.

One notes that, for elliptic orbits, $\sqrt{\zeta} = \Delta E/2$, where ΔE means the increment of the eccentric anomaly over the leapfrog step.

8.5.2 N-Body Formulation with Extrapolation

With the N-body kinetic energy T and force function U,

$$T = \sum_{k=1}^{n}\frac{\mathbf{p}_k^2}{2m_k}, \quad U = \sum_{i<j}\frac{m_i m_j}{r_{ij}}, \quad B = -T(\mathbf{p}(0)) + U(\mathbf{r}(0)),$$

the equations of motion are

$$\mathbf{p}'_k = \frac{1}{U}\frac{\partial U}{\partial \mathbf{r}_k}, \tag{8.61}$$

$$\mathbf{r}'_k = \frac{1}{T+B}\frac{\mathbf{p}_k}{m_k}, \tag{8.62}$$

$$t' = \frac{1}{T+B}. \tag{8.63}$$

Since the derivatives of the coordinates depend only on momenta and those of the momenta depend only on coordinates, it is possible to apply the leapfrog algorithm analogously to the two-body formulation above. For a very close approach of two particles, the motion reduces essentially to two-body motion and thus no catastrophic error will occur. It has been demonstrated (Mikkola and Tanikawa, 1999a,b) that this algorithm can be used successfully even for collision orbits. *Thus a kind of regularization is achieved without coordinate transformation.*

The N-body leapfrog that follows from the above equations of motion,

$$\mathbf{r}_{1/2} = \mathbf{r}_0 + \tfrac{1}{2}h\mathbf{r}'_0, \quad \mathbf{p}_1 = \mathbf{B} + h\mathbf{p}'_{1/2}, \quad \mathbf{r}_1 = \mathbf{r}_{1/2} + \tfrac{1}{2}h\mathbf{r}'_1 \tag{8.64}$$

(where we have dropped the body indices), has a further important feature: due to its time-reversibility, it has the desirable property of having an error expansion that contains only even terms,

$$\delta q = a_2h^2 + a_4h^4 + a_6h^6 + \cdots, \tag{8.65}$$

when integration is done over a fixed interval of time with different (sub)stepsizes h. (Here δq is a generic notation for the errors in the coordinates and velocities.) This makes it possible to apply directly the GBS extrapolation method. One computes the solution over a given interval (step) of s using smaller and smaller substeps h and then extrapolates to zero steplength.

Especially interesting is the fact that the method can be efficient for potentials that differ from the Newtonian $1/r$ behaviour at small distances and that the programming is not more complicated than the writing of a basic N-body code which uses the second-order Bulirsch–Stoer integrator.

In Figure 8.2 is illustrated an interaction of five bodies. One can see that the interaction is strong and complicated. This case was computed using the algorithmic regularization for substeps inside the Bulirsch and Stoer (1966) extrapolation algorithm.

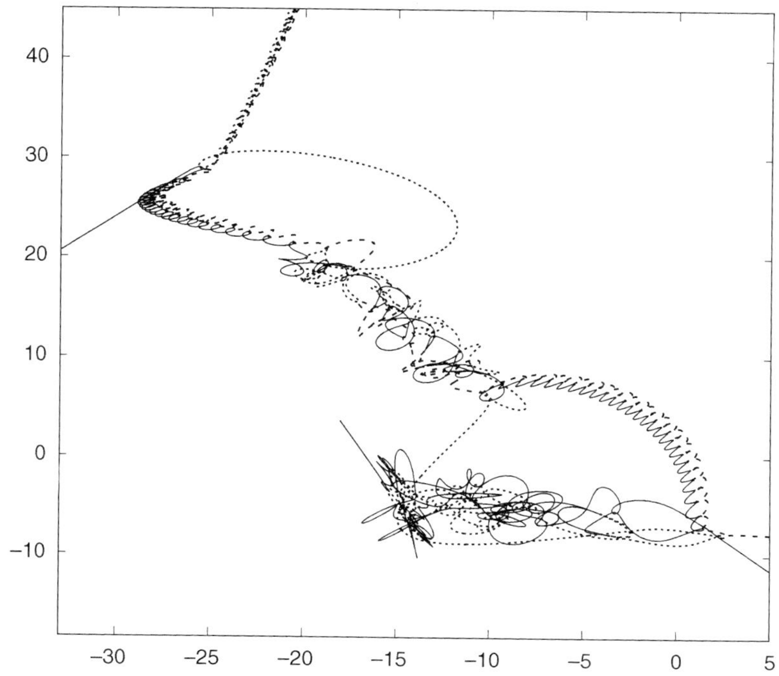

Fig. 8.2. The interaction of a few stars forms typically very complicated orbits before the system disrupts. The final result is often, like here, one hard binary and single bodies.

8.6 Time-Transformed Leapfrog

Here, yet another method to utilize the leapfrog and extrapolation method is considered, for the integration of N-body motions. It is appropriate to begin with a general discussion of second-order differential equations and time transformations.

Consider the special case of a second-order differential equation

$$\ddot{\mathbf{r}} = \mathbf{F}(\mathbf{r}), \tag{8.66}$$

which can be split into the first-order equations

$$\dot{\mathbf{r}} = \mathbf{v}, \quad \dot{\mathbf{v}} = \mathbf{F}(\mathbf{r}). \tag{8.67}$$

These equations can be understood to be the A and B equations (8.20) and (8.21) of a composite method, for

$$\frac{d}{dt}\begin{pmatrix} \mathbf{r} \\ \mathbf{v} \end{pmatrix} = \begin{pmatrix} \mathbf{v} \\ \mathbf{0} \end{pmatrix} + \begin{pmatrix} \mathbf{0} \\ \mathbf{F}(\mathbf{r}) \end{pmatrix}, \tag{8.68}$$

which gives the A equations $\dot{\mathbf{r}} = \mathbf{v}$ and $\dot{\mathbf{v}} = \mathbf{0}$ with the solution $\mathbf{r} = \mathbf{r}(0) + t\mathbf{v}$ and $\mathbf{v} = \mathbf{v}(0)$ and the B equations $\dot{\mathbf{r}} = \mathbf{0}$ and $\dot{\mathbf{v}} = \mathbf{F}(\mathbf{r})$ with the solution $\mathbf{r} = \mathbf{r}(0)$ and $\mathbf{v} = \mathbf{v}(0) + t\mathbf{F}(0)$. When combined in the manner of the ABA algorithm (8.20)–(8.21), these operations give the leapfrog algorithm described in equations (8.2)–(8.4).

Let us next consider a time transformation

$$ds = \Omega(\mathbf{r})\,dt, \tag{8.69}$$

where $\Omega(\mathbf{r})$ is an arbitrary positive function. The use of this in the above system is not trivial, because one often obtains equations (the A and B parts) that are difficult to solve. However, if a new auxiliary variable $W = \Omega$ is introduced, the equations of motion can be written in the form $\mathbf{r}' = \mathbf{v}/W,\ t' = 1/W,\ \mathbf{v}' = \mathbf{F}/\Omega$ (prime means d/ds). There is nothing new in this as yet, but if the value of W can be obtained from the differential equation

$$\dot{W} = \mathbf{v} \cdot \frac{\partial \Omega}{\partial \mathbf{r}} \tag{8.70}$$

(instead of using the relation $W = \Omega$ directly), the equations of motion can be transformed into the form

$$\begin{pmatrix} \mathbf{r}' \\ t' \\ \mathbf{v}' \\ W' \end{pmatrix} = \begin{pmatrix} \mathbf{v}/W \\ 1/W \\ \mathbf{0} \\ 0 \end{pmatrix} + \begin{pmatrix} \mathbf{0} \\ 0 \\ \mathbf{F}(\mathbf{r})/\Omega(\mathbf{r}) \\ \mathbf{v} \cdot \partial \ln(|\Omega(\mathbf{r})|)/\partial \mathbf{r} \end{pmatrix}. \tag{8.71}$$

Thus W is considered to be a variable in the same category as the velocities $\mathbf{v}$. This set of equations is such that its composite parts can be solved, resulting in the time-transformed leapfrog system

$$\mathbf{r}_{1/2} = \mathbf{r}_0 + \frac{h}{2}\frac{\mathbf{v}_0}{W_0}, \tag{8.72}$$

$$t_{1/2} = t_0 + \frac{h}{2}\frac{1}{W_0}, \tag{8.73}$$

$$\mathbf{v}_1 = \mathbf{v}_0 + h\frac{\mathbf{F}(\mathbf{r}_{1/2})}{\Omega(\mathbf{r}_{1/2})}, \tag{8.74}$$

$$W_1 = W_0 + h\frac{\mathbf{v}_0 + \mathbf{v}_1}{2\Omega(\mathbf{r}_{1/2})} \cdot \frac{\partial \Omega(\mathbf{r}_{1/2})}{\partial \mathbf{r}_{1/2}}, \tag{8.75}$$

$$\mathbf{r}_1 = \mathbf{r}_{1/2} + \frac{h}{2}\frac{\mathbf{v}_1}{W_1}, \tag{8.76}$$

$$t_1 = t_{1/2} + \frac{h}{2}\frac{1}{W_1}. \tag{8.77}$$

Combined with the GBS extrapolation method (Gragg, 1964, 1965; Bulirsch and Stoer, 1966; Press et al., 1986, see Chapter 5), this alternative leapfrog can be a powerful integrator for some systems. Note that we have not specified the form of the acceleration $\mathbf{F}$ and/or the time-transformation function Ω and the theory, thus far, can be applied to any system of second-order differential equations that do not contain first derivatives.

8.6.1 Formulation for the Few-Body Problem

We seek the solution of the Newtonian equations of motion

$$\ddot{\mathbf{r}}_k \equiv \mathbf{F}_k = \sum_{j \neq k} m_j \frac{\mathbf{r}_j - \mathbf{r}_k}{r_{kj}^3}, \tag{8.78}$$

where $r_{kj} = |\mathbf{r}_j - \mathbf{r}_k|$. A time-transformation function which does not treat the bodies differently depending on their masses is, for example,

$$\Omega = \sum_{i<j} \frac{\Omega_{ij}}{r_{ij}}, \tag{8.79}$$

where Ω_{ij} are adjustable constants. One may choose, for example, $\Omega_{ij} = 1$ or alternatively $\Omega_{ij} = m_i m_j$ if the masses are comparable. This particular function has the advantage that the analytical solutions for the resulting equations of motion are known to be regular (in the case of collision, the independent variable becomes essentially the eccentric anomaly of the colliding pair). Also, the evaluation of the gradients of Ω can be conveniently done during the computation of the accelerations $\mathbf{F}_k$, for we have

$$\frac{\partial \Omega}{\partial \mathbf{r}_k} \equiv \mathbf{G}_k = \sum_{j \neq k} \Omega_{ij} \frac{\mathbf{r}_j - \mathbf{r}_k}{r_{kj}^3}. \tag{8.80}$$

The time-transformed equations of motion take the form

$$\mathbf{r}'_k = \frac{\mathbf{v}_k}{W}, \quad t' = \frac{1}{W}, \tag{8.81}$$

$$\mathbf{v}'_k = \frac{\mathbf{F}_k}{\Omega}, \quad W' = \frac{1}{\Omega} \sum_k \mathbf{v}_k \cdot \mathbf{G}_k, \tag{8.82}$$

and the construction of the leapfrog algorithm corresponding to (8.72)–(8.77) is immediate.

In the GBS extrapolation method, one first chooses a stepsize, then uses several shorter substeps to integrate over the entire step, and finally there is an extrapolation to zero stepsize. The two common methods to perform the substeps are the midpoint method (for first-order differential equations) and the leapfrog (for second-order equations). Experience has shown that the leapfrog-based algorithm is more powerful and therefore this method, when used for the substeps in the extrapolation procedure, can be expected to be effective in few-body calculations.

Additional Forces

To include external forces one must simply add their contribution to the N-body accelerations $\mathbf{F}_k$ in the above formulae. Let the external accelerations be $\mathbf{f}_k$. The total acceleration to be used in the leapfrog algorithm in place of $\mathbf{F}_k$ is thus $\mathbf{F}_k + \mathbf{f}_k$. If the additional term $\mathbf{f}_k$ depends on velocities, which is the case, for example, for relativistic post-Newtonian (PN) corrections (Section 9.1) (Soffel, 1989; Mora and Will, 2004), the leapfrog algorithm itself must be slightly modified as explained in Mikkola and Merritt (2006, 2008). In this case, equations of the form (omitting the body indices for simplicity)

$$\mathbf{v}' = (\mathbf{F} + \mathbf{f}(\mathbf{v}))/\Omega, \quad W' = \mathbf{v} \cdot \mathbf{G}/\Omega \tag{8.83}$$

replace (8.82). These equations cannot generally be solved accurately. However, if $\mathbf{f}$ is small, a convenient replacement for (8.74) and (8.75) is the implicit midpoint method, for which

$$\mathbf{v}_1 = \mathbf{v}_0 + h(\mathbf{F} + \mathbf{f}(\mathbf{v}_a))/\Omega, \quad \mathbf{v}_a = \frac{\mathbf{v}_0 + \mathbf{v}_1}{2}, \tag{8.84}$$

for the velocity jump, while the W-jump expression (8.75) remains unchanged. (Extrapolation can be used with the implicit midpoint method because it has suitable time symmetry.) The above equation must be solved by iteration, but in practical cases the velocity dependence is very small and thus iteration converges rapidly. Exceptions are extreme relativistic situations, but such events are rare. Another, more complicated, method that solves the velocity dependence problem is described in Section 8.7.

For checking the energy conservation in a simulation with external forces, one must integrate the additional equation for energy change due to the external acceleration

$$E'(\mathbf{v}_1, \cdots, \mathbf{v}_n) = \sum_k m_k \mathbf{v}_k \cdot \mathbf{f}_k(\mathbf{v}_1, \cdots, \mathbf{v}_n)/\Omega, \tag{8.85}$$

which has now been written in a more complete way than the previous equations. This equation can be integrated in the same way as the $\mathbf{v}$ and W equations; i.e. the energy jump is

$$\Delta E = hE'(\mathbf{v}_a), \tag{8.86}$$

where $\mathbf{v}_a$ represents all the average velocities.

Connection with the Logarithmic Hamiltonian

The logarithmic Hamiltonian method (Mikkola and Tanikawa, 1999a,b; Preto and Tremaine, 1999) provides a kind of regularization by the algorithm (without coordinate transformation) as it is exact for the two-body problem. The present algorithm has a close resemblance to the logarithmic Hamiltonian method. If $\Omega = U$ is chosen, e.g. $\Omega_{ij} = m_i m_j$, the methods are, indeed, equivalent.

The logarithmic Hamiltonian equations of motion may be obtained by taking the force function U (negative of potential) of the system as the time-transformation function, i.e. $\Omega = U$, replacing the quantity W by $W = T - E$, where T is the kinetic and E the total energy.

In this case the potential of the system and our time-transformation function are identical, so that

$$\mathbf{F}_k = \partial U/\partial \mathbf{r}_k = \partial \Omega/\partial \mathbf{r}_k = \mathbf{G}_k. \tag{8.87}$$

Then the velocity jump in the leapfrog algorithm may be written as

$$\delta \mathbf{v}_k = h\mathbf{F}_k/\Omega = h\mathbf{G}_k/\Omega, \tag{8.88}$$

and the kinetic energy change becomes

$$\delta T = \sum_k \frac{m_k}{2}(\mathbf{v}_k^2(h) - \mathbf{v}_k^2(0)) = \sum_k \frac{m_k}{2}(\mathbf{v}_k(h) + \mathbf{v}_k(0)) \cdot \delta \mathbf{v}_k. \tag{8.89}$$

By (8.88), this equals the change in the value of W as expressed by (8.75) in a slightly different notation. Because $W(0) = \Omega(0)$ it is now clear that, in the case of equal masses, at any moment $W = T - E$. Thus the equivalence of the two methods for the case $\Omega = U$ has been demonstrated.

8.6.2 Algorithmic Regularization Chain

Energies can be written as

$$T = \sum_{k=1}^{N} \frac{m_k}{2}\mathbf{v}_k^2, \quad U = \sum_{i<j\le N} \frac{m_i m_j}{r_{ij}}. \tag{8.90}$$

The main purpose of the chain structure (in this method) is to reduce the effect of roundoff error. If we define

$$\mathbf{X}_k = \mathbf{r}_{i_k} - \mathbf{r}_{j_k}, \quad \mathbf{V}_k = \mathbf{v}_{i_k} - \mathbf{v}_{j_k}, \tag{8.91}$$

and

$$\mathbf{X} = (\mathbf{X}_1, \mathbf{X}_2, \cdots, \mathbf{X}_{N-1}), \quad \mathbf{V} = (\mathbf{V}_1, \mathbf{V}_2, \cdots, \mathbf{V}_{N-1}), \tag{8.92}$$

then the equations of motion can be written

$$\dot{\mathbf{X}} = \mathbf{V}, \quad \dot{\mathbf{V}} = \mathbf{A}(\mathbf{X}) + \mathbf{f}, \tag{8.93}$$

where $\mathbf{A}$ is the N-body acceleration and $\mathbf{f}$ some external acceleration (other bodies, relativistic terms, etc.).

One may use the two equivalent time transformations (Mikkola and Merritt, 2006)

$$ds = [\alpha(T + B) + \beta\omega + \gamma]dt = [\alpha U + \beta\Omega + \gamma]dt, \tag{8.94}$$

where α, β and γ are adjustable constants, and Ω is a function of the coordinates. Coordinate equations are

$$t' = 1/(\alpha(T + B) + \beta\omega + \gamma), \tag{8.95}$$

$$\mathbf{X}' = t'\mathbf{V}. \tag{8.96}$$

Velocity equations are

$$\tilde{t}' = 1/(\alpha U + \beta\Omega + \gamma), \tag{8.97}$$

$$\mathbf{V}' = \tilde{t}'\ (\mathbf{A} + \mathbf{f}), \tag{8.98}$$

$$\omega' = \tilde{t}'\ \frac{\partial\Omega}{\partial\mathbf{X}} \cdot \mathbf{V}, \tag{8.99}$$

$$B' = -\tilde{t}'\ \frac{\partial T}{\partial\mathbf{V}} \cdot f. \tag{8.100}$$

The leapfrog for the chain vectors **X** and **V** can be written as the two mappings

$\mathbf{X}(s)$:

$$\delta t = s/(\alpha(T + B) + \beta\omega + \gamma), \tag{8.101}$$

$$t = t + \delta t, \tag{8.102}$$

$$\mathbf{X} \to \mathbf{X} + \delta t\mathbf{V}, \tag{8.103}$$

$\mathbf{V}(s)$:

$$\widetilde{\delta t} = s/(\alpha U + \beta\Omega + \gamma), \tag{8.104}$$

$$\mathbf{V} \to \mathbf{V} + \widetilde{\delta t}(\mathbf{A} + \langle\mathbf{f}\rangle), \tag{8.105}$$

$$B \to B + \widetilde{\delta t}\left\langle\frac{\partial T}{\partial\mathbf{V}}\right\rangle \cdot \langle\mathbf{f}\rangle, \tag{8.106}$$

$$\omega \to \omega + \widetilde{\delta t}\frac{\partial\Omega}{\partial\mathbf{X}} \cdot \langle\mathbf{V}\rangle, \tag{8.107}$$

where $\langle\mathbf{f}\rangle$, $\langle\partial/\partial\mathbf{V}\rangle$ and $\langle\mathbf{V}\rangle$ are the averages over the advancement of $\mathbf{V}$.

8.6.3 Details

Finding and Updating the Chain

First we find the shortest interparticle vector which is adopted as the first part of the chain. The chain is then refilled by adding to it the relative vector from the particle to the nearest end of the already known chain (Mikkola and Aarseth, 1993). When all particles are included, they are renumbered along the chain as $1, 2, \ldots, N$ for ease of programming.

To reduce roundoff problems the transformation from the old chain vectors $\mathbf{X}_k$ to the new ones is done directly by expressing the new chain vectors as sums of the old ones as in Mikkola and Aarseth (1993).

Transformations

When the particles are named along the chain as $1, 2, \ldots, N$, one can evaluate

$$\mathbf{X}_k = \mathbf{r}_{k+1} - \mathbf{r}_k, \tag{8.108}$$

$$\mathbf{V}_k = \mathbf{v}_{k+1} - \mathbf{v}_k. \tag{8.109}$$

The centre-of-mass quantities are

$$M = \sum_k m_k, \tag{8.110}$$

$$\mathbf{r}_{cm} = \sum_k m_k \mathbf{r}_k / M, \tag{8.111}$$

$$\mathbf{v}_{cm} = \sum_k m_k \mathbf{v}_k / M. \tag{8.112}$$

The inverse transformation is done by simple summation,

$$\widetilde{\mathbf{r}}_1 = \mathbf{0}, \tag{8.113}$$

$$\widetilde{\mathbf{v}}_1 = \mathbf{0}, \tag{8.114}$$

$$\widetilde{\mathbf{r}}_{k+1} = \widetilde{\mathbf{r}}_k + \mathbf{X}_k, \tag{8.115}$$

$$\widetilde{\mathbf{v}}_{k+1} = \widetilde{\mathbf{v}}_k + \mathbf{V}_k, \tag{8.116}$$

followed by reduction to the centre of mass,

$$\widetilde{\mathbf{r}}_{cm} = \sum_k m_k \widetilde{\mathbf{r}}_k / M, \tag{8.117}$$

$$\widetilde{\mathbf{v}}_{cm} = \sum_k m_k \widetilde{\mathbf{v}}_k / M, \tag{8.118}$$

$$\mathbf{r}_k = \widetilde{\mathbf{r}}_k - \widetilde{\mathbf{r}}_{cm}, \tag{8.119}$$

$$\mathbf{v}_k = \widetilde{\mathbf{v}}_k - \widetilde{\mathbf{v}}_{cm}. \tag{8.120}$$

Note that it is not always necessary to reduce the coordinates to the centre-of-mass system since accelerations only depend on the differences.

Equations of Motion and the Leapfrog

One writes the equations of motion as

$$\dot{\mathbf{X}}_k = \mathbf{V}_k, \tag{8.121}$$

$$\dot{\mathbf{V}}_k = \mathbf{F}_{k+1} - \mathbf{F}_k + \mathbf{f}_{k+1} - \mathbf{f}_k, \tag{8.122}$$

where $\mathbf{f}_k$ are the individual external accelerations and the N-body accelerations $\mathbf{F}_k$ are

$$\mathbf{F}_k = -\sum_{j \neq k} m_j \frac{\mathbf{r}_{jk}}{|\mathbf{r}_{jk}|^3}, \tag{8.123}$$

where, for $j < k$,

$$\mathbf{r}_{jk} = \begin{cases} \mathbf{r}_k - \mathbf{r}_j, & \text{if } k > j+2, \\ \mathbf{X}_j, & \text{if } k = j+1, \\ \mathbf{X}_j + \mathbf{X}_{j+1}, & \text{if } k = j+2. \end{cases} \tag{8.124}$$

For $k > j$ one uses $\mathbf{r}_{jk} = -\mathbf{r}_{kj}$. In the acceleration computation the use of $\mathbf{X}_j$ and $\mathbf{X}_j + \mathbf{X}_{j+1}$ reduces the roundoff effect significantly. This is one of the most important points in the algorithm.

The kinetic energy is evaluated as usual as

$$T = \frac{1}{2} \sum_k m_k \mathbf{v}_k^2, \tag{8.125}$$

while the potential,

$$U = \sum_{i<j} \frac{m_i m_j}{|\mathbf{r}_{ij}|}, \tag{8.126}$$

is obtained along with the accelerations according to (8.124). For the time-transformation function Ω it seems advantageous to use

$$\Omega = \sum_{i<j} \frac{\Omega_{ij}}{|\mathbf{r}_{ij}|}. \tag{8.127}$$

Here Ω_{ij} are adjustable constants (see below).

Now one is able to evaluate the two time-transformation functions

$$t' = 1/(\alpha(T+B) + \beta\omega + \gamma), \tag{8.128}$$

$$\tilde{t}' = 1/(\alpha U + \beta\Omega + \gamma), \tag{8.129}$$

which are equivalent along the correct solution, i.e. $t' = \tilde{t}'$. The evolution of ω, with $\omega(0) = \Omega(0)$, is obtained by

$$\dot{\omega} = \sum_k \frac{\partial \Omega}{\partial \mathbf{r}_k} \cdot \mathbf{v}_k. \tag{8.130}$$

In the presence of external perturbations the binding energy evolves according to

$$\dot{B} = -\sum_k m_k \mathbf{v}_k \cdot \mathbf{f}_k. \tag{8.131}$$

The leapfrog for the chain vectors $\mathbf{X}_k$ and $\mathbf{V}_k$ can be written as the two mappings

$\mathbf{X}(s)$:

$$\delta t = s/(\alpha(T+B) + \beta\omega + \gamma), \tag{8.132}$$

$$t = t + \delta t, \tag{8.133}$$

$$\mathbf{X}_k \to \mathbf{X}_k + \delta t \mathbf{V}_k, \tag{8.134}$$

$\mathbf{V}(s)$:

$$\widetilde{\delta t} = s/(\alpha U + \beta\Omega + \gamma), \tag{8.135}$$

$$\mathbf{V}_k \to \mathbf{V}_k + \widetilde{\delta t}(\mathbf{F}_{k+1} - \mathbf{F}_k + \mathbf{f}_{k+1} - \mathbf{f}_k), \tag{8.136}$$

$$B \to B + \widetilde{\delta t} \sum_k \left(-m_k \langle \mathbf{v}_k \rangle \cdot \mathbf{f}_k\right), \tag{8.137}$$

$$\omega \to \omega + \widetilde{\delta t} \sum_k \frac{\partial \Omega}{\partial \mathbf{r}_k} \cdot \langle \mathbf{v}_k \rangle, \tag{8.138}$$

where $\langle \mathbf{v}_k \rangle$ is the average of the initial and final $\mathbf{v}$ here (obtained from the $\mathbf{V}$ according to the equations in Section 8.6.3). It is necessary to evaluate the individual velocities $\mathbf{v}_k$ because the expressions for B' and ω' in terms of the chain vector velocities $\mathbf{V}_k$ are rather cumbersome.

A leapfrog step can be written as

$$\mathbf{X}(h/2)\mathbf{V}(h)\mathbf{X}(h/2)$$

and a sequence of n steps as

$$\mathbf{X}(h/2)\left[\prod_{\nu=1}^{n-1}(\mathbf{V}(h)\mathbf{X}(h))\right]\mathbf{V}(h)\mathbf{X}(h/2). \tag{8.139}$$

This is useful with the extrapolation method when advancing the system over a total time interval of length $= nh$.

8.6.4 Time-Transformation Alternatives

If one takes

$$\Omega_{ij} = m_i m_j, \tag{8.140}$$

then $\alpha = 0$, $\beta = 1$, $\gamma = 0$ is mathematically equivalent to $\alpha = 1$, $\beta = \gamma = 0$ as was shown originally in Mikkola and Aarseth (2002), and here in Section 8.6.1. However, numerically these are not equivalent, mainly due to roundoff errors in updating the value of Ω, and the logH alternative is numerically more stable. However, for proper treatment of small bodies, some function Ω is to be used.

For increased numerical stability for motions of the big bodies and smoothing the encounters of small bodies, the recipe is: $\alpha = 1$, $\beta \neq 0$ (but small) and

$$\Omega_{ij} = \begin{cases} \widetilde{m}^2, & \text{if } m_i * m_j < \epsilon\, \widetilde{m}^2, \\ 0, & \text{otherwise,} \end{cases} \tag{8.141}$$

where $\widetilde{m}^2 = \sum_{i<j} m_i m_j/(N(N-1)/2)$ is the mean mass product and $\epsilon \sim 10^{-3}$. It may be advisable to integrate the equation (8.130) for ω even if $\beta = 0$, just to make the integrator (extrapolation method!) use short steps if $\dot{\omega}$ is large, thus giving higher precision when required.

8.7 Velocity-Dependent Perturbations

In many dynamics problems the extrapolation method (Gragg, 1965; Bulirsch and Stoer, 1966; Press et al., 1986) is one of the most efficient high-precision integration methods. In the perturbed two-body problem, or more generally in the N-body problem with external perturbations, such as relativistic post-Newtonian terms (Section 9.1), one has velocity-dependent perturbations.

This section suggests a new simple idea for constructing an explicit time-symmetrical algorithmic regularization method for the substeps required in any extrapolation algorithm, even when the forces depend on velocities. These ideas were published by Hellström and Mikkola (2010).

8.7.1 The Basic Idea

Consider first the simple second-order equation of motion

$$\ddot{\mathbf{r}} = \mathbf{A}(\mathbf{r}, \dot{\mathbf{r}}), \tag{8.142}$$

which, in terms of the velocity $\mathbf{v} = \dot{\mathbf{r}}$, may be rewritten as

$$\dot{\mathbf{r}} = \mathbf{v}, \tag{8.143}$$

$$\dot{\mathbf{v}} = \mathbf{A}(\mathbf{r}, \mathbf{v}). \tag{8.144}$$

Because the expression for $\dot{\mathbf{v}}$ depends on $\mathbf{v}$, it is not possible to construct an explicit leapfrog algorithm for this system. It has been customary to suggest the use of the implicit midpoint method (see e.g. Mikkola and Aarseth, 2002; Mikkola and Merritt, 2006)

$$\mathbf{r}_{1/2} = \mathbf{r}_0 + \frac{h}{2}\mathbf{v}_0, \tag{8.145}$$

$$\mathbf{v}_1 = \mathbf{v}_0 + h\mathbf{A}(\mathbf{r}_{1/2}, (\mathbf{v}_0 + \mathbf{v}_1)/2), \tag{8.146}$$

$$\mathbf{r}_1 = \mathbf{r}_{1/2} + \frac{h}{2}\mathbf{v}_1, \tag{8.147}$$

where the value of $\mathbf{v}_1$ must be obtained by iteration. Thus this method can be computationally expensive.

The new idea consists of introducing an auxiliary velocity $\mathbf{w}$ and writing down the equations of motion in the extended space of variables $\mathbf{r}$, $\mathbf{v}$, $\mathbf{w}$,

$$\dot{\mathbf{r}} = \mathbf{v}, \tag{8.148}$$

$$\dot{\mathbf{w}} = \mathbf{A}(\mathbf{r}, \mathbf{v}), \tag{8.149}$$

$$\dot{\mathbf{v}} = \mathbf{A}(\mathbf{r}, \mathbf{w}), \tag{8.150}$$

where the right-hand sides do not depend on left-hand side variables and consequently a leapfrog-like explicit algorithm can be constructed. To proceed, in n steps, over a pre-given time interval of length $\tau = nh$, one sets the initial value $\mathbf{w}(0) = \mathbf{v}(0)$ and carries out n leapfrog steps. One such step can be written as

$$\mathbf{r}_{1/2} = \mathbf{r}_0 + \frac{h}{2}\mathbf{v}_0, \tag{8.151}$$

$$\mathbf{w}_{1/2} = \mathbf{w}_0 + \frac{h}{2}\mathbf{A}(\mathbf{r}_{1/2}, \mathbf{v}_0), \tag{8.152}$$

$$\mathbf{v}_1 = \mathbf{v}_0 + h\mathbf{A}(\mathbf{r}_{1/2}, \mathbf{w}_{1/2}), \tag{8.153}$$

$$\mathbf{w}_1 = \mathbf{w}_{1/2} + \frac{h}{2}\mathbf{A}(\mathbf{r}_{1/2}, \mathbf{v}_1), \tag{8.154}$$

$$\mathbf{r}_1 = \mathbf{r}_{1/2} + \frac{h}{2}\mathbf{v}_1, \tag{8.155}$$

which is a completely explicit and time-symmetrical algorithm in the extended parameter space. Consequently this algorithm can also be efficiently used as a substep method in any extrapolation algorithm (e.g. Gragg, 1965; Bulirsch and Stoer, 1966; Press et al., 1986).

One should note that in the above algorithms the 'unperturbed' part $\mathbf{A}(\mathbf{r})$ must be evaluated only once per leapfrog step. In the implicit midpoint equation (8.146) the number of iterations necessary for convergence to full precision (which is necessary for the extrapolation method to work) is generally at least three, often considerably larger. The new algorithm evaluates the perturbation exactly three times per leapfrog step and thus is often more efficient than the implicit midpoint method. One must, however, note that the exact properties of these methods in applications depend in a complicated way on the installation and the 'interaction' of the extrapolation routine and stepsize selection algorithm with the leapfrog routine.

8.7.2 Perturbed Two-Body Problem

In this section we introduce the algorithmic regularization modification of our method. This leapfrog formulation has the property that, if the perturbation acceleration $\mathbf{a} = 0$, then for the two-body problem the result is a correct trajectory with only an $O(h^2)$ time error.

The basic equation we consider reads

$$\ddot{\mathbf{r}} = -m\mathbf{r}/r^3 + \mathbf{a}(\mathbf{r}, \mathbf{v}, t). \tag{8.156}$$

To introduce the time-transformed equations of the algorithmic regularization method we define

$$T = \tfrac{1}{2}\mathbf{v}^2, \tag{8.157}$$

$$U = m/r, \tag{8.158}$$

$$B = U - T \quad \text{(binding energy)}, \tag{8.159}$$

$$\mathbf{A} = \frac{\partial U(\mathbf{r})}{\partial \mathbf{r}} = -\frac{m\mathbf{r}}{r^3}. \tag{8.160}$$

Using for short the notation

$$T_e = T + B \ \ (= U), \tag{8.161}$$

one may write

$$t' = 1/T_e \ \ (= 1/U), \tag{8.162}$$

$$\mathbf{r}' = \mathbf{v}/T_e, \tag{8.163}$$

$$B' = -\mathbf{v} \cdot \mathbf{a}(\mathbf{r}, \mathbf{v}, t)/U, \tag{8.164}$$

$$\mathbf{v}' = (\mathbf{A}(\mathbf{r}, t) + \mathbf{a}(\mathbf{r}, \mathbf{v}, t))/U, \tag{8.165}$$

and the new idea involves introducing the auxiliary velocity $\mathbf{w}$ (actually $\mathbf{w} = \mathbf{v}$), which allows the equations of motion

$$t' = 1/T_e \ (= 1/U), \tag{8.166}$$

$$\mathbf{r}' = \mathbf{v}/T_e, \tag{8.167}$$

$$\mathbf{w}' = (\mathbf{A}(\mathbf{r}) + \mathbf{a}(\mathbf{r}, \mathbf{v}, t))/U, \tag{8.168}$$

$$\mathbf{v}' = (\mathbf{A}(\mathbf{r}) + \mathbf{a}(\mathbf{r}, \mathbf{w}, t))/U, \tag{8.169}$$

$$B' = -\mathbf{v} \cdot \mathbf{a}(\mathbf{r}, \mathbf{v}, t)/U, \tag{8.170}$$

in which the derivative of any quantity does not explicitly depend on the quantity itself.

At this point it is convenient to define separate operations (rather routines) $\mathbf{X}(h_s)$ for moving the coordinates and $\mathbf{V}(h_s)$ for the velocity changes. The $\mathbf{X}(h_s)$ routine reads

$\mathbf{X}(h_s)$:

$$T_e = \mathbf{p}^2/2 + B, \tag{8.171}$$

$$\delta t = h_s/T_e, \tag{8.172}$$

$$t \to t + \delta t, \tag{8.173}$$

$$\mathbf{r} \to \mathbf{r} + \delta t \mathbf{v}, \tag{8.174}$$

and the $\mathbf{V}(h_s)$ routine is

$\mathbf{V}(h_s)$:

$$U = m/r, \tag{8.175}$$

$$\widetilde{\delta t} = h_s/U, \tag{8.176}$$

$$\mathbf{w} \to \mathbf{w} + \frac{\widetilde{\delta t}}{2}(\mathbf{A}(\mathbf{r}) + \mathbf{a}(\mathbf{r}, \mathbf{v}, t)), \tag{8.177}$$

$$\mathbf{v} \to \mathbf{v} + \widetilde{\delta t}(\mathbf{A}(\mathbf{r}) + \mathbf{a}(\mathbf{r}, \mathbf{w}, t)), \tag{8.178}$$

$$B \to B - \widetilde{\delta t}\mathbf{w} \cdot \mathbf{a}(\mathbf{r}, \mathbf{w}, t), \tag{8.179}$$

$$\mathbf{w} \to \mathbf{w} + \frac{\widetilde{\delta t}}{2}(\mathbf{A}(\mathbf{r}) + \mathbf{a}(\mathbf{r}, \mathbf{v}, t)). \tag{8.180}$$

With the routines $\mathbf{X}$ and $\mathbf{V}$ one can symbolize a leapfrog step sequence over a time interval $\tau = nh$ as

$$\mathbf{X}(h/2)\,[\,\mathbf{V}(h)\mathbf{X}(h)\,]^{n-1}\,\mathbf{V}(h)\mathbf{X}(h/2),$$

which can be understood as a subroutine calling sequence in which one first takes a half-step with $\mathbf{X}(h/2)$ followed by $n-1$ alternate calls of routines $\mathbf{V}(h)$ and $\mathbf{X}(h)$ and finally one must do one more evaluation of $\mathbf{V}(h)$ followed by the final half-step $\mathbf{X}(h/2)$.

8.7.3 Some Simple Extensions

1. Additional variables (e.g. BH spin)
   $\dot{\mathbf{s}} = \mathbf{f}(\mathbf{r}, \mathbf{v}, t, \mathbf{s})$
   and
   $\mathbf{a} = \mathbf{a}(\mathbf{r}, \mathbf{v}, t, \mathbf{s})$
   can be added as 'extra components' of $\mathbf{v}$ and $\mathbf{w}$ vectors, but here more complicated versions of equations are required.
2. N-body problem with similar external perturbations.

8.8 Special Cases

In this section we discuss some particular systems. In simple cases we only present the division of the Hamiltonian into the two parts which can then be used according to the instructions given in the previous section. Some of the notation is according to (8.57)–(8.60).

8.8.1 The Quasi-Keplerian Leapfrog

The quasi-Keplerian problem has the Hamiltonian

$$H = \frac{1}{2}\mathbf{p}^2 - \frac{M}{r} - \frac{\alpha}{r^2}. \tag{8.181}$$

This may be useful, e.g. for inclusion of the relativistic precession (Hellström and Mikkola, 2009) and also in connection of some symplectic methods with time transformation (Mikkola, 1997). To use the logH formulation (with $B = -H$) one possibility is to write

$$\Lambda = \log\left(\frac{1}{2}\mathbf{p}^2 + B\right) - \log\left(\frac{M}{r} + \frac{\alpha}{r^2}\right), \tag{8.182}$$

but, somewhat surprisingly, the other alternative of writing

$$\Lambda = \log\left(\frac{1}{2}\mathbf{p}^2 + B - \frac{\alpha}{r^2}\right) - \log\left(\frac{M}{r}\right) \tag{8.183}$$

is more beneficial: it gives exactly the correct trajectory!

An illustrative example is given in Figure 8.3. Like in the case of the Newtonian two-body problem, we have here only the time error. This has to do with the fact that, in polar coordinates (r, θ) and momenta (p_r, p_θ), one can write

$$H_0 = \frac{1}{2}\mathbf{p}^2 - \frac{\alpha}{r^2} = \frac{1}{2}\left(p_r^2 + \frac{p_\theta^2 - 2\alpha}{r^2}\right), \tag{8.184}$$

and the form of the Hamiltonian is similar to that in a two-body problem. More precisely, if we introduce a new angle ψ by transforming the momentum as

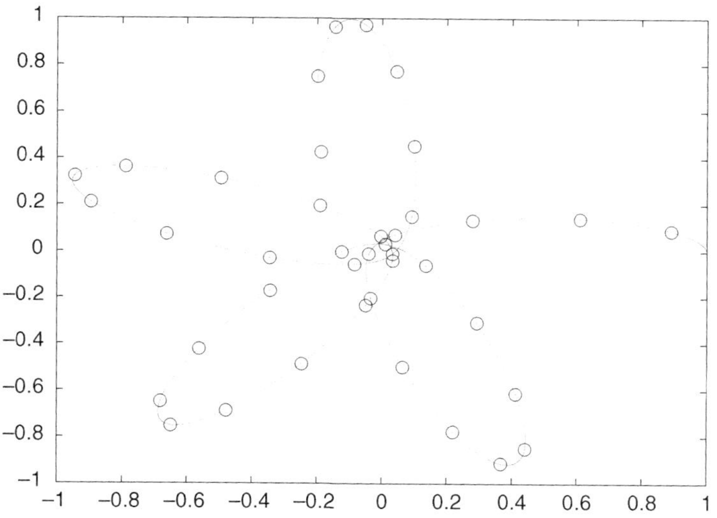

Fig. 8.3. Quasi-Keplerian motion computed with the generalized leapfrog using two different steps. The trajectories are independent of the stepsize (checked also numerically to full precision using multiple h_s steps according to (8.58)).

$$p_\psi^2 = p_\theta^2 - 2\alpha, \tag{8.185}$$

then the generating function

$$S = \psi\sqrt{p_\theta^2 - 2\alpha} \tag{8.186}$$

gives the canonical transformation

$$\theta = \frac{\partial S}{\partial p_\theta} = \psi p_\theta/\sqrt{p_\theta^2 - 2\alpha} = \psi/\sqrt{1 - 2\alpha/p_\theta^2}. \tag{8.187}$$

From this we see that the angular speeds in the two systems (θ, p_θ) and (ψ, p_ψ) differ by a constant (of motion) factor. The Hamiltonian in the new system reads

$$\Lambda = \log\left(\frac{1}{2}p_r^2 + \frac{1}{2}\frac{p_\psi^2}{r^2} + B\right) - \log\left(\frac{M}{r}\right), \tag{8.188}$$

and this is clearly a two-body Hamiltonian in polar coordinates. Thus the logarithmic leapfrog gives correct results in the (r, ψ) space because it works for the two-body problem. (Note that the variables used in the computations are immaterial.) After solving the motion in the (r, ψ) space, one can obtain the increment of θ by using the relation

$$\Delta\theta = \Delta\psi/\sqrt{1 - 2\alpha/p_\theta^2}, \tag{8.189}$$

to get the coordinates and velocities in the original system. Because the angular increments differ only by a constant factor we can conclude that the solution for the quasi-Keplerian problem is also correct with the logarithmic leapfrog (except for the time error).

Advancing the System Using the Kinetic Energy

The angular velocity in the rotating ellipse is $\dot{\psi} = p_\psi/r^2$ instead of $\dot{\theta} = p_\theta/r^2$. The general formulation for the motion can be done using the increment of the rotation angle $\Delta\theta$ and the areal velocity vector $\mathbf{p}_\theta = \mathbf{r}_0 \times \mathbf{v}_0$,

$$\mathbf{r} = \frac{r}{r_0}\left(\cos(\Delta\theta)\mathbf{r}_0 + \frac{\sin(\Delta\theta)}{p_\theta}\mathbf{p}_\theta \times \mathbf{r}_0\right), \quad \mathbf{v} = \frac{\dot{r}}{r}\mathbf{r} + \frac{1}{r^2}\mathbf{p}_\theta \times \mathbf{r}, \tag{8.190}$$

which is correct for any central motion (0 subscript means beginning of step). Note that in this system the velocity $\mathbf{v} = \dot{\mathbf{r}}$ and momentum $\mathbf{p}$ are the same and in these equations we use them alternatively. For the time corrections, as well as for the stepsize parameter h in terms of h_s, the same formulae (8.57)–(8.59) as in the normal two-body case apply here. This is because the time transformation is still $t' = 1/(\frac{1}{2}\mathbf{p}^2 + B - \alpha/r^2) = r/M$ and the motion of the angle does not explicitly affect the time (except by the value of the energy).

The derivation of the equations of motion is simple since the Hamiltonian that we need,

$$H_K = \mathbf{B}^2/2 - \alpha/r_0^2 = \tfrac{1}{2}p_r^2 + \tfrac{1}{2}p_\psi^2/r^2, \tag{8.191}$$

is (effectively) the kinetic energy and so r^2 is quadratic in time. In fact it is easy to derive the equation of motion

$$\frac{d^2(r^2)}{dt^2} = 4H_K, \tag{8.192}$$

and so r^2 is a quadratic function of time because H_K is here a constant (during this operation). The increment of θ is then easily obtained by the integral

$$\Delta\theta = p_\theta \int dt/r^2, \tag{8.193}$$

which results in an arctan expression (or the corresponding hyperbolic expression depending on the quadratic expression for r^2). After obtaining h as a function of h_s according to (8.58) one proceeds a half-step using

$$H_K = \mathbf{v}_0^2/2 - \alpha/r_0^2, \quad \mathbf{p}_\theta = \mathbf{r}_0 \times \mathbf{v}_0, \quad p_\psi^2 = p_\theta^2 - 2\alpha, \tag{8.194}$$

$$\delta t = \tfrac{1}{2}h/(H_K + B), \tag{8.195}$$

$$\eta = \mathbf{r}_0 \cdot \mathbf{v}_0, \tag{8.196}$$

$$r^2 = r_0^2 + \delta t(2\eta + 2H_K\delta t), \tag{8.197}$$

$$d_\theta = \delta t/(r^2 + \eta\delta t), \tag{8.198}$$

$$\Delta\theta = p_\theta\, d_\theta\, \mathrm{at}_1(p_\psi^2 d_\theta^2), \tag{8.199}$$

$$t \to t + \delta t, \tag{8.200}$$

and uses these results in the equations (8.190) for $\mathbf{r}$ and $\mathbf{v}$. Above, the special function $\mathrm{at}_1(z)$ is defined as

$$\begin{aligned}\mathrm{at}_1(z) &= \mathrm{atan}(\sqrt{z})/\sqrt{z} = \mathrm{atanh}(\sqrt{-z})/\sqrt{-z} \\ &= 1 - z/3 + z^2/5 - z^3/7 + \cdots,\end{aligned} \tag{8.201}$$

where the trigonometric form is used for $z > 0$ and the hyperbolic one when $z < 0$, while for small values of z a Padé approximant may be faster.

Momentum Jump

The solution for the quasi-Keplerian problem was formulated in a different way in Mikkola (1997) and Hellström and Mikkola (2009), but the advantage of the leapfrog algorithm is that one can add another perturbation $R(t, \mathbf{r})$ easily by replacing the second term in Λ by

$$-\log(M/r + R(t, \mathbf{r})),$$

in which case the momentum jumps take the form

$$\mathbf{p} \to \mathbf{p} + h\left(-M\mathbf{r}/r^3 + \frac{\partial R(t, \mathbf{r})}{\partial \mathbf{r}}\right)/(M/r + R(t, \mathbf{r})), \tag{8.202}$$

$$B \to B + h\frac{\partial R(t, \mathbf{r})}{\partial t}/(M/r + R(t, \mathbf{r})). \tag{8.203}$$

If R does not depend explicitly on time, then naturally B remains a constant.

8.8.2 Comets and the Galactic Tide

The Galactic tidal field affects the motion of the very-long-period comets. The tidal potential can be approximated by a quadratic expression. In addition one must include the rotation of the Galaxy. Thus the Hamiltonian takes the form

$$H = \tfrac{1}{2}\mathbf{p}^2 - M/r + g_{xx}x^2 + g_{yy}y^2 + g_{zz}z^2 - n(xp_y - yp_x), \tag{8.204}$$

where the g are the tidal field coefficients and n the angular speed of the Galaxy. Here the perturbing part is very small near the Sun, but not necessarily so in the outer parts of the Oort Cloud. A convenient logH formulation is

$$\Lambda = \log(\tfrac{1}{2}\mathbf{p}^2 + g_{xx}(x^2 - y^2) + g_{zz}z^2 - n(xp_y - yp_x) + B) - \log(M/r), \tag{8.205}$$

where we have used the fact that $g_{yy} = -g_{xx}$. The argument of the first logarithm is quadratic in the coordinates and momenta and thus equations of motion are easily solvable, while the second part simply gives momentum jumps ($= -h\mathbf{r}/r^2$). Thus a generalized leapfrog can be constructed according to the general guidelines given in Section 8.2. If the distance r is very large then the $-\log(M/r)$ term gives only a small perturbation to the motion defined by the Galactic tidal term. At the same time the timestep can be quite large, but this is not a problem since the tidal contribution is integrated analytically. Also the (recommended) use of the extrapolation method anyway takes care of any accuracy problem.

8.8.3 Binary Asteroids

The case of a binary asteroid (in the point-mass approximation) is in some respects similar to the comets in the Galactic field. Approximating the motion of the binary asteroids' centre of mass by an ellipse and using the tidal field approximation we can write the equations of motion for the relative motion of the asteroid pair as

$$\ddot{\mathbf{q}} = -M\left(\frac{\mathbf{q}}{r^3} - 3\frac{(\mathbf{r}\cdot\mathbf{q})\mathbf{r}}{r^5}\right) - (m_1 + m_2)\frac{\mathbf{q}}{q^3}, \tag{8.206}$$

where $\mathbf{r}$ is the heliocentric distance of the binary in the elliptic motion around the Sun (taken to be a known function of time), $\mathbf{q}$ is the relative position vector of the components, M is the mass of the Sun, and m_1 and m_2 are the masses of the asteroids. The above equation of motion can be derived from the Hamiltonian

$$H = \tfrac{1}{2}\mathbf{p}^2 + M(\tfrac{1}{2}r^2\mathbf{q}^2 - \tfrac{3}{2}(\mathbf{r}\cdot\mathbf{q})^2)/r^5 - (m_1 + m_2)/q, \tag{8.207}$$

where $\mathbf{p}$ is the conjugate momentum of $\mathbf{q}$ (in this formulation $\mathbf{p}$ is equal to the relative velocity of the binary components). The corresponding logH takes the form

$$\begin{aligned}\Lambda &= \log(\tfrac{1}{2}\mathbf{p}^2 + M(\tfrac{1}{2}r^2\mathbf{q}^2 - \tfrac{3}{2}(\mathbf{r}\cdot\mathbf{q})^2)/r^5 + B) \\ &\quad - \log\left(\frac{m_1 + m_2}{q}\right),\end{aligned} \tag{8.208}$$

where $B = -H$ (numerically) is the momentum of time. The two-body motion of the binary centre of mass around the Sun satisfies the normal equation of motion

$$\ddot{\mathbf{r}} = -M\frac{\mathbf{r}}{r^3}. \tag{8.209}$$

If we use the notation $\mathbf{q}$ for the variation of $\mathbf{r}$, the corresponding variational equation reads

$$\ddot{\mathbf{q}} = -M\left(\frac{\mathbf{q}}{r^3} - 3\frac{(\mathbf{r}\cdot\mathbf{q})\mathbf{r}}{r^5}\right), \tag{8.210}$$

which is equal to what the Hamiltonian inside the first logarithm in equation (8.208) gives for $\mathbf{q}$ motion. Thus the advancements of the vectors $\mathbf{q}$ and $\mathbf{p}$, when the first part of the logH is applied, can be obtained as solutions for the two-body variational equations. Detailed algorithms for analytical computation are listed in several papers (e.g. Mikkola and Innanen, 1999). Since the vector $\mathbf{r}$ is time-dependent, the value of B changes during every step. The equation reads $\dot{B} = -\partial H_0/\partial t$ but because $\dot{H}_0 = \partial H_0/\partial t$ we see that the jump in the value of B equals the jump in the value of $-H_0$ and thus it can be obtained easily as

$$B_{new} = B_{old} + (H_{0\ old} - H_{0\ new}), \tag{8.211}$$

where the subscripts *old* and *new* mean the values in the beginning and end of the step.

In the case of circular motion of $\mathbf{r}$ in units in which $M = 1$, $r = 1$ and in the rotating coordinate system, one can alternatively use the formulation (with subscripts x and y indicating the components)

$$\Lambda = \log(\tfrac{1}{2}\mathbf{p}^2 + \tfrac{1}{2}(\mathbf{q}^2 - 3q_x^2) - (q_x p_y - q_y p_x) + B) - \log\left(\frac{m_1 + m_2}{q}\right). \tag{8.212}$$

This is more easy to use since now H_0 is a quadratic polynomial in coordinates and momenta. Again, in between the H_0 steps, the jumps in the momenta are simply

$$\mathbf{p} \to \mathbf{p} - h\mathbf{q}/q^2. \tag{8.213}$$

8.9 Tidal Friction and Spin of Stars

Tidal Friction and Quadrupolar Distortion

In Kiseleva, Eggleton and Mikkola (1998) a simple theory of tidal friction between stars was published. In addition to tidal friction (TF) the quadrupolar distortion (QD) was treated. For two stars the force on star 1 due to star 2 is the following:

$$\mathbf{r}_{12} = \mathbf{r}_1 - \mathbf{r}_2, \quad \mathbf{v}_{12} = \dot{\mathbf{r}}_1 - \dot{\mathbf{r}}_2, \tag{8.214}$$

$$\mathbf{F}_{12} = -\left[\frac{m_1 m_2}{r_{12}^3} + \frac{6(m_2^2 A_1 + m_1^2 A_2)}{r_{12}^8}\right]\mathbf{r}_{12} + \left[\frac{27}{2}\frac{\sigma_1 m_2^2 A_1^2 + \sigma_2 m_1^2 A_2^2}{r_{12}^{10}}\mathbf{r}_{12}\cdot\mathbf{v}_{12}\right]\mathbf{r}_{12}, \tag{8.215}$$

$$\sigma_i = \frac{\lambda_i}{Q_i^2 m_i R_i^2}\sqrt{\frac{m_1}{R_i^3}}, \quad A_i = \frac{R_i^5 Q_i}{1 - Q_i}. \tag{8.216}$$

The expression $\mathbf{F}_{12}$ includes the effects of QD and TF, to be added to the Newtonian acceleration of $\mathbf{r}_{12}$. The quantities R_i are the stellar radii and the coefficients λ_i are the dimensionless dissipation rates for the stars 1 and 2 (these indices can really be any numbers). Often the numerical value is estimated to be something like $\lambda \approx 10^{-4}$ or 10^{-5}, but it is not known accurately. The quantity Q_i is a dimensionless measure of the distortability of the star and can be approximated as

$$Q_i = \frac{8}{5}\frac{\int_0^{m_i} r^5 dm_i}{R_i^5 m_i} \approx \frac{3}{5}\left(1 - \frac{n}{5}\right)^{2.215} \exp(0.0245n - 0.096n^2 - 0.0084n^3). \tag{8.217}$$

This expression is an interpolation approximation for stars that have polytropic structure with the index $= n$. The index n is in the interval $0 \leq n < 5$ and it depends on the stars' evolutionary phase.

Effect of Stellar Rotation

The rotation of stars gives additional terms in equation (8.215). Those terms can be written as $\delta\mathbf{F}_1 + \delta\mathbf{F}_2$ where

$$\delta F_1 = \frac{9}{2}\frac{\sigma_1 m_2^2 A_1^2}{r_{12}^{10}}\left(\mathbf{r}_{12} \times \mathbf{v}_{12} - \Omega_1 r_{12}^2\right) \times \mathbf{r}_{12}, \tag{8.218}$$

$$\frac{d}{dt}(I_1\Omega_1) = -\mathbf{r}_{12} \times \delta\mathbf{F}_1. \tag{8.219}$$

Here I_1 is the moment of inertia of star 1 and Ω_1 is the angular velocity (spin) of the star. In these equations one can simply interchange the indices 1 and 2 to get results for star 2. Since these expressions depend on velocities, the integration methods discussed in Section 8.7 must be used.

8.10 An Efficient Three-Body Algorithm

The method discussed in this section can also be found in Mikkola and Tanikawa (2013a).

8.10.1 Logarithmic Hamiltonian Method

The logarithmic Hamiltonian method was first published by Mikkola and Tanikawa (1999a) and Preto and Tremaine (1999). The Hamiltonian is

$$H = T(\mathbf{p}) - U(\mathbf{q}, t). \tag{8.220}$$

The vector $\mathbf{p}$ is the momentum, $\mathbf{q}$ is the coordinate vector, t is the time, T is the kinetic energy and U is the force function (negative of potential energy). One may define the momentum of time $B = -H$. The logarithmic Hamiltonian

$$\Lambda = \log(T + B) - \log(U) \tag{8.221}$$

gives the equations of motion

$$t' = \frac{\partial \Lambda}{\partial B} = 1/(T + B), \tag{8.222}$$

$$\mathbf{q}' = \frac{\partial \Lambda}{\partial \mathbf{p}} = \frac{\partial T}{\partial \mathbf{p}}/(T + B), \tag{8.223}$$

$$B' = \frac{\partial \Lambda}{\partial t} = \frac{\partial U}{\partial t}/U, \tag{8.224}$$

$$\mathbf{p}' = \frac{\partial \Lambda}{\partial \mathbf{q}} = \frac{\partial U}{\partial \mathbf{q}}/U. \tag{8.225}$$

It is possible to use the simple leapfrog algorithm for approximately solving the equations. With the simplified notation $\mathbf{X} = (t, \mathbf{q})$ and $\mathbf{W} = (B, \mathbf{p})$ one can write a single leapfrog step

$$\begin{aligned} \mathbf{X}_{1/2} &= \mathbf{X}_0 + \tfrac{1}{2}h\mathbf{X}'(\mathbf{W}_0), \\ \mathbf{W}_1 &= \mathbf{W}_0 + h\mathbf{W}'(\mathbf{X}_{1/2}), \\ \mathbf{X}_1 &= \mathbf{X}_{1/2} + \tfrac{1}{2}h\mathbf{X}'(\mathbf{W}_1). \end{aligned} \tag{8.226}$$

Here h is the stepsize and indices 0 and 1 mean values at the beginning and end of the step.

This algorithm gives the exact trajectory for the two-body problem and more generally regular results even in the case of point-mass collision orbits. Thus the leapfrog can be used to carry out the substeps necessary in the Bulirsch–Stoer method (Bulirsch and Stoer, 1966).

A standard experimental case is the Pythagorean problem, which was solved accurately for the first time using the KS-transformation (Szebehely and Peters, 1967). With the present method this example is easily done even using leapfrog alone.

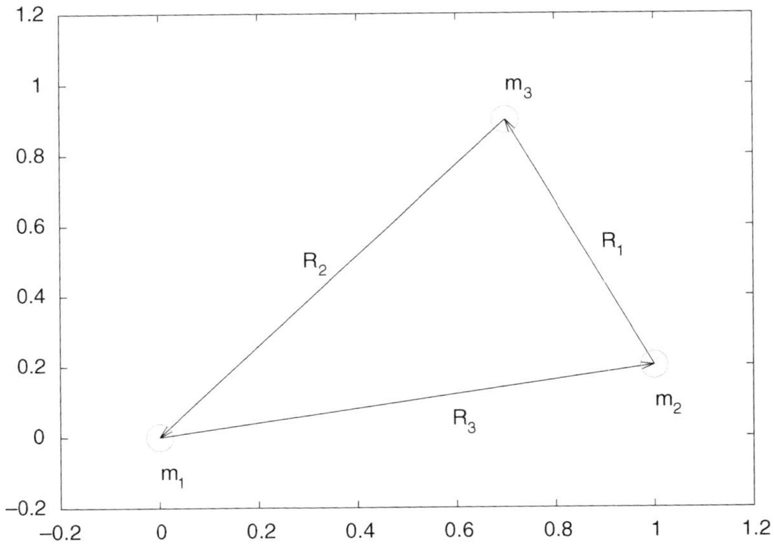

Fig. 8.4. Labelling of the relative coordinates.

8.11 Formulation with Relative Coordinates

One can conveniently use the relative position vectors (see Figure 8.4)

$$\mathbf{R}_1 = \mathbf{r}_3 - \mathbf{r}_2, \ \ \mathbf{R}_2 = \mathbf{r}_1 - \mathbf{r}_3, \ \ \mathbf{R}_3 = \mathbf{r}_2 - \mathbf{r}_1 \tag{8.227}$$

as new coordinates and, instead of momenta, the relative velocities $\mathbf{V}_k = \dot{\mathbf{R}}_k$, which are formed in the same way as the relative coordinates. With the total mass $M = \sum_k m_k$ and the new coordinates and velocities, the energies T and U can be written as

$$T = \frac{1}{2M}(m_2 m_3 \mathbf{V}_1^2 + m_1 m_3 \mathbf{V}_2^2 + m_1 m_2 \mathbf{V}_3^2), \tag{8.228}$$

$$U = m_2 m_3 / R_1 + m_1 m_3 / R_2 + m_1 m_2 / R_3. \tag{8.229}$$

Using the Newtonian equations of motion one obtains for the relative coordinates the following:

$$\dot{\mathbf{R}}_k = \mathbf{V}_k, \ \ \dot{\mathbf{V}}_k = -M \frac{\mathbf{R}_k}{|\mathbf{R}_k|^3} + m_k \sum_\nu \frac{\mathbf{R}_\nu}{|\mathbf{R}_\nu|^3}. \tag{8.230}$$

After the application of the time transformations they read

$$t' = 1/(T + B), \ \ \mathbf{R}'_k = \mathbf{V}_k/(T + B), \ \ \mathbf{V}'_k = \dot{\mathbf{V}}_k/U, \tag{8.231}$$

which are suitable for the leapfrog algorithm (8.226).

8.12 Algorithms

It is convenient to compute the quantities

$$\mu_k = m_i m_j, \tag{8.232}$$

where the indices (k, i, j) are $(1, 2, 3)$ in some order (but all different). Thus $\mu_1 = m_2 m_3$, $\mu_2 = m_1 m_3$, $\mu_3 = m_1 m_2$. One also needs the total mass $M = m_1 + m_2 + m_3$.

8.12.1 Initial Values

First one needs the relative coordinates $\mathbf{R}_k$ and velocities $\mathbf{V}_k$:

$$\mathbf{R}_1 = \mathbf{r}_3 - \mathbf{r}_2, \quad \mathbf{R}_2 = \mathbf{r}_1 - \mathbf{r}_3, \quad \mathbf{R}_3 = \mathbf{r}_2 - \mathbf{r}_1, \tag{8.233}$$

$$\mathbf{V}_1 = \mathbf{v}_3 - \mathbf{v}_2, \quad \mathbf{V}_2 = \mathbf{v}_1 - \mathbf{v}_3, \quad \mathbf{V}_3 = \mathbf{v}_2 - \mathbf{v}_1. \tag{8.234}$$

Then the initial value of centre-of-mass energies and the binding energy B are obtained by

$$T = \frac{1}{2M} \sum_k \mu_k \mathbf{V}_k^2, \tag{8.235}$$

$$U = \sum_k \mu_k / R_k, \tag{8.236}$$

$$B = U - T. \tag{8.237}$$

8.12.2 Leapfrogging

1. Moving coordinates with the stepsize h_s using the operation $\mathbf{X}(h_s)$:

$$T_B = \frac{1}{2M} \sum_k \mu_k \mathbf{V}_k^2 + B, \tag{8.238}$$

$$dt = h_s / T_B, \tag{8.239}$$

$$t \to t + dt, \tag{8.240}$$

$$\mathbf{R}_k \to \mathbf{R}_k + dt\, \mathbf{V}_k \quad (k = 1, 2, 3). \tag{8.241}$$

2. Moving velocities with stepsize h_s using the operation $\mathbf{V}(h_s)$:

$$U = M \sum_k \mu_k / R_k, \tag{8.242}$$

$$\mathbf{S} = \sum_k \mathbf{R}_k / R_k^3, \tag{8.243}$$

$$\delta t = h_s / U, \tag{8.244}$$

$$\mathbf{V}_k \to \mathbf{V}_k + \delta t(-M \mathbf{R}_k / R_k^3 + m_k \mathbf{S}), \quad (k = 1, 2, 3). \tag{8.245}$$

3. Leapfrog over long intervals: with the operations $\mathbf{X}(h_s)$ and $\mathbf{V}(h_s)$ it is possible to compute over n steps, so that the total step has length $= nh$, as

$$\mathbf{X}(h/2)[\mathbf{V}(h)\mathbf{X}(h)]^{n-1}\mathbf{V}(h)\mathbf{X}(h/2), \tag{8.246}$$

where the power $(n-1)$ indicates repetition of the operations.

8.12.3 Triangle Corrections

Due to the choice of the elements we have the constants

$$\sum_k \mathbf{R}_k = \mathbf{0}, \quad \sum_k \mathbf{V}_k = \mathbf{0}. \tag{8.247}$$

In practice one can integrate all three coordinates and velocities but quite often carry out a triangle correction.

First one finds the longest vector $\mathbf{R}_k$, and recomputes it, and the corresponding velocity, as

$$\mathbf{R}_k = -(\mathbf{R}_i + \mathbf{R}_j) \quad (i,j,k \text{ different}), \tag{8.248}$$

$$\mathbf{V}_k = -(\mathbf{V}_i + \mathbf{V}_j). \tag{8.249}$$

Numerical tests have shown that this operation is near-optimal.

8.12.4 Transforming to Inertial Coordinates

A simple operation to transform the relative coordinates and velocities to centre-of-mass quanties reads

$$\mathbf{r}_1 = (m_3\mathbf{R}_2 - m_2\mathbf{R}_3)/M, \tag{8.250}$$

$$\mathbf{r}_2 = (m_1\mathbf{R}_3 - m_3\mathbf{R}_1)/M, \tag{8.251}$$

$$\mathbf{r}_3 = (m_2\mathbf{R}_1 - m_1\mathbf{R}_2)/M, \tag{8.252}$$

$$\mathbf{v}_1 = (m_3\mathbf{V}_2 - m_2\mathbf{V}_3)/M, \tag{8.253}$$

$$\mathbf{v}_2 = (m_1\mathbf{V}_3 - m_3\mathbf{V}_1)/M, \tag{8.254}$$

$$\mathbf{v}_3 = (m_2\mathbf{V}_1 - m_1\mathbf{V}_2)/M. \tag{8.255}$$

Since in the three-body dynamics the timescales of phenomena vary by large factors, a constant-stepsize leapfrog in the integration is not satisfactory. For accurate results it is often necessary to use the leapfrog only for substeps in an extrapolation method. The Bulirsch–Stoer (Bulirsch and Stoer, 1966) extrapolation for this purpose works very well. The same may be true also for the polynomial extrapolation methods of Gragg (1965).

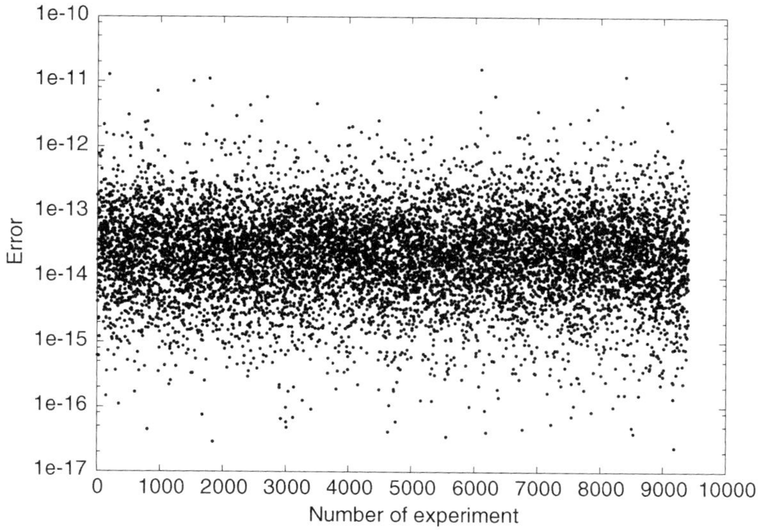

Fig. 8.5. Energy errors/potential in more than 9000 three-body experiments after evolution to the time of ejection of one of the masses, which leaves behind a binary.

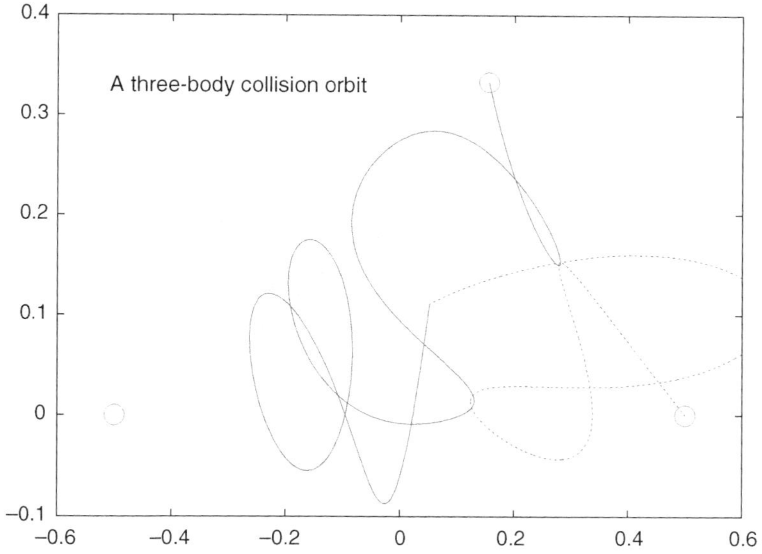

Fig. 8.6. A three-body collision orbit in a two-dimensional system. The initial velocities of the particles were here $= 0$, which means that the orbit is one in the so-called free-fall three-body problem. The bodies collide at about the middle of the figure, where the orbital curves arrive at the same point, so that the bodies obviously form an equilateral triangle just before the collision, which is the mathematical end of these orbits.

8.12.5 A Numerical Test

Years ago almost 10 000 three-body simulations starting from random initial conditions were done by Mikkola and Tanikawa (2007). The energy errors in those experiments, at the point when the systems had disrupted, are shown in Figure 8.5. These are actually the values of the logarithmic Hamiltonian, which is the main constant of motion in the system where the Hamiltonian is the logarithmic one. In a sense this is also a kind of relativistic error $= \delta E/U$. As the scatter of errors shows, these are essentially values that one may expect from roundoff errors alone.

In Figure 8.6 is plotted a three-body collision orbit. It was computed using a code based on the method explained in this section. The result was obtained, among other similar orbits, in connection to the work of Tanikawa et al. (2019). These orbits are not easy to find, but the method is fast and accurate, which made it possible.

Chapter 9
Motion in the Field of a Black Hole

9.1 Post-Newtonian Terms

The post-Newtonian (PN) terms that can be used to approximate the effect of a black hole (BH) have been published by many authors. Here the ones by Mora and Will (2004) (up to order 3.5PN, i.e. to terms that are divided by c^7) are listed. These are more complete than the often used (up to order 2.5PN) terms (Soffel, 1989), but in some cases the 3PN and 3.5PN terms seem to give unexpected effects. Thus it may be better to restrict to order 2.5PN anyway (Dey et al., 2018). With these terms it may be more efficient to use the explicit method explained near the end of Section 8.7.1 instead of the implicit one (Section 8.6.1), when the results are obtained by an extrapolation method. An even more detailed explanation is given in Section 8.7.2. The explicit formulation is more efficient because these terms are rather complicated to evaluate and thus the iterations required by the implicit midpoint method would be time-consuming. If both components are black holes the equations can still be used for both of them, at least in a first approximation.

In the following r is the distance between the BH and a star and v is the relative velocity:

$$m = m_1 + m_2, \tag{9.1}$$

$$\eta = m_1 m_2 / m^2, \tag{9.2}$$

$$\mathbf{n} = \frac{\mathbf{r}}{r}, \tag{9.3}$$

$$A_1 = 2(2+\eta)\left(\frac{m}{r}\right) - (1+3\eta)v^2 + 1.5\eta\dot{r}^2, \tag{9.4}$$

$$
\begin{aligned}
A_2 = &-\frac{3}{4}(12+29\eta)\left(\frac{m}{r}\right)^2 - \eta(3-4\eta)v^4 \\
&- \frac{15}{8}\eta(1-3\eta)\dot{r}^4 + 0.5\eta(13-4\eta)\left(\frac{m}{r}\right)v^2 \\
&+ (2+25\eta+2\eta^2)\left(\frac{m}{r}\right)\dot{r}^2 + 1.5\eta(3-4\eta)v^2\dot{r}^2, \qquad (9.5)
\end{aligned}
$$

$$
A_{2.5} = \frac{8}{5}\eta\left(\frac{m}{r}\right)\dot{r}\left(\frac{17}{3}\frac{m}{r} + 3v^2\right), \qquad (9.6)
$$

$$
\begin{aligned}
A_3 = &\left[16 + \left(\frac{1399}{12} - \frac{41}{16}\pi^2\right)\eta + \frac{71}{2}\eta^2\right]\left(\frac{m}{r}\right)^3 \\
&+ \eta\left(\frac{20827}{840} + \frac{123}{64}\pi^2 - \eta^2\right)\left(\frac{m}{r}\right)^2 v^2 \\
&- \left[1 + \left(\frac{22717}{168} + \frac{615}{64}\pi^2\right)\eta + \frac{11}{8}\eta^2 - 7\eta^3\right]\left(\frac{m}{r}\right)^2\dot{r}^2 \\
&- \frac{1}{4}\eta(11-49\eta+52\eta^2)v^6 + \frac{35}{16}\eta(1-5\eta+5\eta^2)\dot{r}^6 \\
&- \frac{1}{4}\eta(75+32\eta-40\eta^2)\left(\frac{m}{r}\right)v^4 \\
&- \frac{1}{2}\eta(158-69\eta-60\eta^2)\left(\frac{m}{r}\right)\dot{r}^4 \\
&+ \eta(121-16\eta-20\eta^2)\left(\frac{m}{r}\right)v^2\dot{r}^2 \\
&+ \frac{3}{8}\eta(20-79\eta+60\eta^2)v^4\dot{r}^2 - \frac{15}{8}\eta(4-18\eta+17\eta^2)v^2\dot{r}^4, \\
&\qquad (9.7)
\end{aligned}
$$

$$
\begin{aligned}
A_{3.5} = &-\frac{8}{5}\eta\left(\frac{m}{r}\right)\dot{r}\left[\frac{23}{14}(43+14\eta)\left(\frac{m}{r}\right)^2 + \frac{3}{28}(61+70\eta)v^4 + 70\dot{r}^4\right. \\
&+ \frac{1}{42}(519-1267\eta)\left(\frac{m}{r}\right)v^20 + 0.25(147+188\eta)\left(\frac{m}{r}\right)\dot{r}^2 \\
&\left.- \frac{15}{4}(19+2\eta)v^2\dot{r}^2\right], \qquad (9.8)
\end{aligned}
$$

$$
B_1 = 2(2-\eta)\dot{r}, \qquad (9.9)
$$

$$
\begin{aligned}
B_2 = &-\frac{1}{2}\dot{r}\left[(4+41\eta+8\eta^2)\left(\frac{m}{r}\right)\right. \\
&\left.- \eta(15+4\eta)v^2 + 3\eta(3+2\eta)\dot{r}^2\right], \qquad (9.10)
\end{aligned}
$$

$$
B_{2.5} = -\frac{8}{5}\eta\left(\frac{m}{r}\right)\left[3\left(\frac{m}{r}\right) + v^2\right], \qquad (9.11)
$$

$$
\begin{aligned}
B_3 = \dot{r}\Bigg\{&\left[4 + \left(\frac{5849}{840} + \frac{123}{32}\pi^2\right)\eta - 25\eta^2 - 8\eta^3\right]\left(\frac{m}{r}\right)^2 \\
&+ \frac{1}{8}\eta(65-152\eta-48\eta^2)v^4 + \frac{15}{8}\eta(3-8\eta-2\eta^2)\dot{r}^4
\end{aligned}
$$

$$+ \eta(15 + 27\eta + 10\eta^2)\left(\frac{m}{r}\right)v^2$$
$$- \frac{1}{6}\eta(329 + 177\eta + 108\eta^2)\left(\frac{m}{r}\right)\dot{r}^2$$
$$- \frac{3}{4}\eta(16 - 37\eta - 16\eta^2)v^2\dot{r}^2\Bigg\}, \tag{9.12}$$

$$B_{3.5} = \frac{8}{5}\eta\left(\frac{m}{r}\right)\left[\frac{1}{42}(1325 + 546\eta)\left(\frac{m}{r}\right)^2\right.$$
$$+ \frac{1}{28}(313 + 42\eta)v^{2^2} + 75\dot{r}^4 - \frac{1}{42}(205 + 777\eta)\left(\frac{m}{r}\right)v^2$$
$$\left. + \frac{1}{12}(205 + 424\eta)\left(\frac{m}{r}\right)\dot{r}^2 - \frac{3}{4}(113 + 2\eta)v^2\dot{r}^2\right]. \tag{9.13}$$

The total effect on the acceleration is now given by

$$A_{tot} = A_1/c^2 + A_2/c^4 + A_{2.5}/c^5 + A_3/c^6 + A_{3.5}/c^7, \tag{9.14}$$
$$B_{tot} = B_1/c^2 + B_2/c^4 + B_{2.5}/c^5 + B_3/c^6 + B_{3.5}/c^7, \tag{9.15}$$
$$\ddot{\mathbf{r}} = \ddot{\mathbf{r}}_N - \frac{m}{r^2}(A_{tot}\mathbf{n} + B_{tot}\mathbf{v}) + \ddot{\mathbf{x}}_{so} + \ddot{\mathbf{x}}_Q. \tag{9.16}$$

In the formula (9.16) for $\ddot{\mathbf{r}}$ the term $\ddot{\mathbf{r}}_N$ is the Newtonian acceleration and the formulae for the spin–orbit terms $\ddot{\mathbf{x}}_{so}$ and $\ddot{\mathbf{x}}_Q$ are given below.

9.1.1 Spin–Orbit Terms

The formulae in this section are mostly from Valtonen et al. (2010). The leading-order spin–orbit term $\ddot{\mathbf{x}}_{so}$ reads

$$\ddot{\mathbf{x}}_{so} = \frac{m^2}{c^3 r^3}\frac{1 + \sqrt{1 - 4\eta}}{4} \times \mathcal{X}\{[12\mathbf{s}_1 \cdot \mathbf{n} \times \mathbf{v}]\mathbf{n}$$
$$+ [(9 + 3\sqrt{1 - 4\eta})\dot{r}](\mathbf{n} \times \mathbf{s}_1) - [7 + \sqrt{1 - 4\eta}](\mathbf{v} \times \mathbf{s}_1)\}, \tag{9.17}$$

where the Kerr parameter $\mathcal{X} = |\vec{\alpha}|$ (see the $\vec{\alpha}$ below) and $\mathbf{s}_1 = \vec{\alpha}/|\vec{\alpha}|$ is the unit vector defining the direction of the rotation axes of the BH.

The quadrupole–monopole term $\ddot{\mathbf{x}}_Q$ to be added to $\ddot{\mathbf{r}}$ reads

$$\ddot{\mathbf{x}}_Q = \mathcal{X}^2 \frac{3m_1^2 m}{2c^2 r^4}\{[5(\mathbf{n} \cdot \mathbf{s}_1)^2 - 1]\mathbf{n} - 2(\mathbf{n} \cdot \mathbf{s}_1)\mathbf{s}_1\}, \tag{9.18}$$

and the equation of motion for $\mathbf{s}_1$ can be written

$$\dot{\mathbf{s}}_1 = \Omega \times \mathbf{s}_1, \tag{9.19}$$

where

$$\Omega = \frac{m\eta}{2c^2r^3}\frac{7+\sqrt{1-4\eta}}{1+\sqrt{1-4\eta}}(\mathbf{n}\times\mathbf{v}). \tag{9.20}$$

Merger of Bodies with a Black Hole

If another body comes too close to a BH, they merge. The distance, the so-called Schwarzschild radius $r = R_s$, is

$$R_s = 2GM/c^2, \tag{9.21}$$

where G is the gravitational constant (in this book mostly units with $G = 1$ are used). It is a coordinate singularity in the solutions of Einstein's equation, but also a distance at which the bodies merge. Radius R_s is also called the *event horizon* because nothing inside that distance can come out of the BH. In Newtonian physics that distance looks like the one at which the escape velocity of any particle equals c, the speed of light. In numerical integrations, the R_s distance is often too small for efficient computation and therefore it may be reasonable to merge the bodies somewhat earlier.

A Problem at the Event Horizon

One other thing that is usually not mentioned is that if one uses, for example, only the 1PN terms and allows the bodies to come to the event horizon distance, they simply reflect instead of merging. The reason is mathematical, not physical. With one term more in the PN terms the bodies really dive together and thus merge. A numerically computed example is illustrated in Figure 9.1.

Gravitational Radiation Effect

As an example of the effect of the relativistic gravitational radiation terms, Figure 9.2 can be presented. One can see that the radiation affects the in-spiral of bodies near a black hole. The terms were taken into account up to the 2.5PN term. This system is a fictitious one with BH mass $= 1$, and the in-spiralling body mass $= 0.001$. (Black holes of mass $= 1$ may not exist but mathematically these numbers can be used.) As the figure tells, the relative orbit is almost a straight line and during the closest approach to the BH the pericentre precesses significantly and the system loses energy so that the semi-major axis reduces quickly. One should note here that the smallest body mass cannot be set to $= 0$ because zero mass does not cause gravitational radiation. The starting apocentre distance in this experiment was $= 1$ and the eccentricity of the initial relative orbit was set to be $e = 1 - 10^{-6}$.

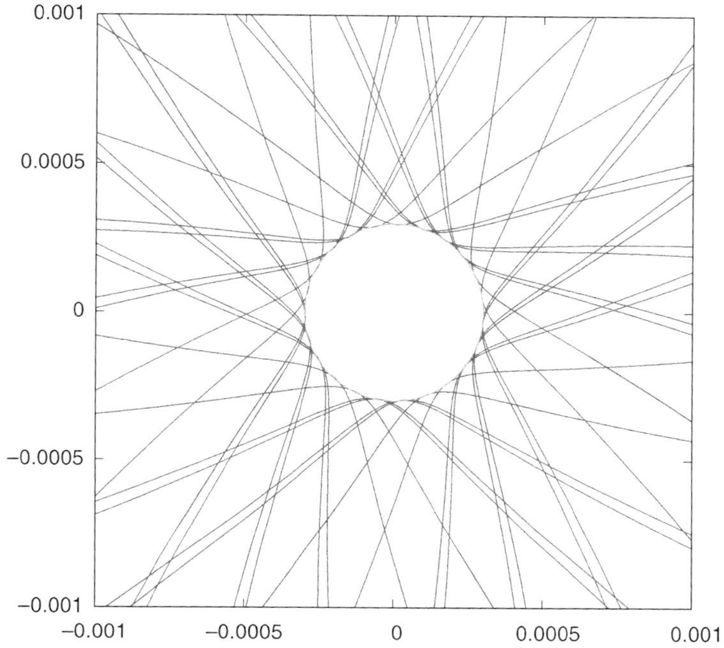

Fig. 9.1. If only the 1PN terms are used, for example, then a body reflects from the event horizon. This is not physical, but is a mathematical effect due to the incompleteness of the expressions.

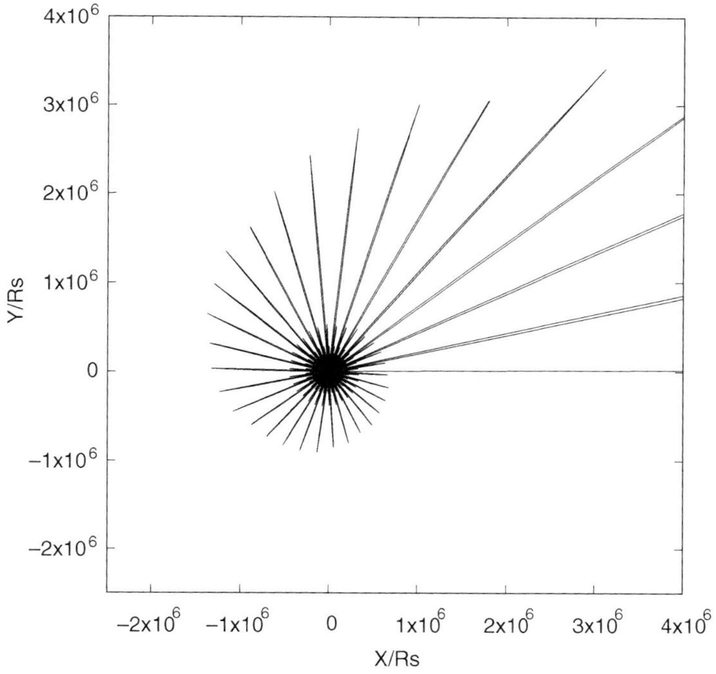

Fig. 9.2. This figure demonstrates the in-spiralling effect of gravitational radiation when a smaller body comes very close to a black hole. The lengths are measured in units of the black hole radius $R_s = 2M/c^2$ (in the system with $G = 1$).

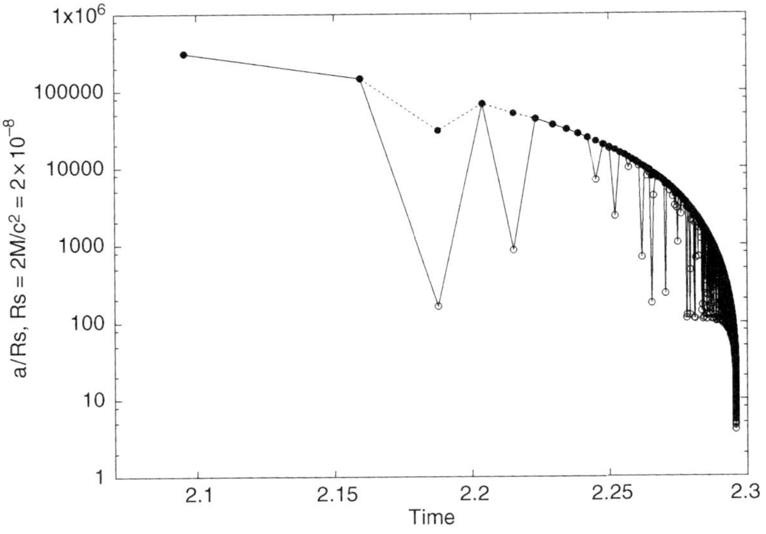

Fig. 9.3. Semi-major axes computed with the Newtonian formula (lower fluctuating part of the plot) and the PN-corrected value (at the upper bound of the curves). The X-axis is the time and the Y-axis is the semi-major axis measured in terms of the black hole radius.

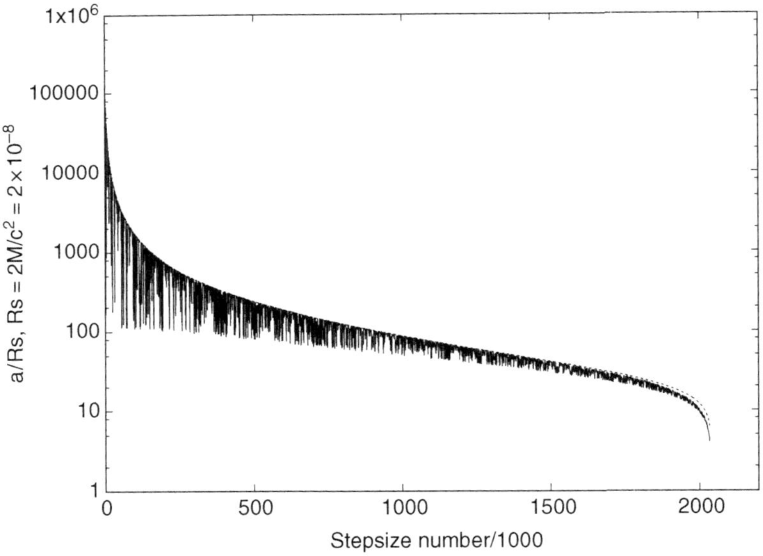

Fig. 9.4. The same quantity as in Figure 9.3, but the X-axis units are measured in terms of thousands of integration steps with the AR leapfrog and BS integrator.

The in-spiralling time before merger of the bodies was $= 6.494$ times the (formal) initial period computed according to the Newtonian formulae. In Figures 9.3 and 9.4 the evolution of the semi-major axis in this experiment is illustrated. One can see the big difference of the values computed by the Newtonian expressions and those with PN correction. The PN-corrected values give the elements at the apocentre where velocities are almost zeros. Since the Y-axis in those figures is logarithmic, the differences between uncorrected and corrected values is actually large. Thus the Newtonian orbital elements are not always even approximately correct.

9.1.2 PN-Corrected Orbital Elements

It is possible to find expressions for orbital elements (at least for energy and angular momentum, i.e. a and e). The expressions are available in Blanchet and Iyer (2003) and Mora and Will (2004), for example:

$$M = M_1 + M_2, \tag{9.22}$$

$$\eta = M_1 M_2 / M^2, \tag{9.23}$$

$$E_0 = \tfrac{1}{2} v^2 - M/r, \tag{9.24}$$

$$\mathbf{j} = \mathbf{r} \times \mathbf{v}, \tag{9.25}$$

$$E_1 = \frac{1}{2}\left(\frac{M}{r}\right)^2 + \frac{3}{8}(1 - 3\eta)v^4 + \frac{1}{2}((3+\eta)v^2 + \eta\dot{r}^2)\frac{M}{r}, \tag{9.26}$$

$$C_1 = (3+\eta)\frac{M}{r} + \frac{1}{2}(1 - 3\eta)v^2, \tag{9.27}$$

$$\begin{aligned} E_2 = & -\frac{1}{4}(2 + 15\eta)\frac{M^3}{r} + \frac{5}{16}(1 - 7\eta + 13\eta^2)v^6 \\ & + \frac{1}{8}(14 - 55\eta + 4\eta^2)\frac{M^2}{r}v^2 + \frac{1}{8}(4 + 69\eta + 12\eta^2)\frac{M^2}{r}\dot{r}^2 \\ & + \frac{1}{8}(21 - 23\eta - 27\eta^2)\frac{M}{r}v^4 \\ & + \frac{1}{4}\eta(1 - 15\eta)\frac{M}{r}v^2\dot{r}^2 - \frac{3}{8}\eta(1 - 3\eta)\frac{M}{r}\dot{r}^4, \end{aligned} \tag{9.28}$$

$$\begin{aligned} C_2 = & \frac{1}{4}(14 - 41\eta + 4\eta^2)\frac{M^2}{r} + \frac{3}{8}(1 - 7\eta + 13\eta^2)v^4 \\ & + \frac{1}{2}(7 - 10\eta - 9\eta^2)\frac{M}{r}v^2 \frac{1}{2}(2 + 5\eta)\frac{M}{r}\dot{r}^2, \end{aligned} \tag{9.29}$$

$$E_{PN} = E_0 + E_1/c^2 + E_2/c^4, \tag{9.30}$$

$$\mathbf{J}_{PN} = \mathbf{j}(1 + C_1/c^2 + C_2/c^4), \tag{9.31}$$

$$a_{PN} = -M/(2E_{PN}), \tag{9.32}$$

$$e_{PN}^2 = 1 + \frac{1}{2}\left(\frac{\mathbf{J}_{PN}}{M}\right)^2 E_{PN}. \tag{9.33}$$

Above the expressions with the subscripts PN mean the PN-corrected values.

9.1.3 Black Hole Spins

Two black holes, both rotating, have angular momenta that interact with the orbital motion and each other's rotation. In addition to the normal PN terms (Soffel, 1989; Mora and Will, 2004), the terms due to the spinning of the BH (i.e. Kerr holes) should be included. Normally, if the big BH (m_1) is much larger than the smaller one (m_2), the spin of m_2 may often be neglected in numerical integration, but here the necessary equations are anyway given. This section is mostly from Gopakumar (2009, private communication).

The notation

$$\mathbf{r}_{ij} = \mathbf{x}_i - \mathbf{x}_j, \quad r_{ij} = |\mathbf{r}_{ij}|, \quad \mathbf{n}_{ij} = \mathbf{r}_{ij}/r_{ij}, \quad \mathbf{v}_{ij} = \mathbf{v}_i - \mathbf{v}_j \tag{9.34}$$

is used, where $\mathbf{x}$ are position vectors and $\mathbf{v}$ the velocities given in terms of the relative BH spins,

$$\vec{\alpha} = \mathbf{S}/S_{max}, \; S_{max} = Gm^2/c, \tag{9.35}$$

where $\mathbf{S}$ is the spin (angular momentum) of the BH. Clearly the length of the $\vec{\alpha}$ is < 1. It tells the BH's rotation axis direction and the fraction of the spin of the maximal value. One gets the acceleration of the body m_i as (Will, 2009, private communication)

$$\begin{aligned}\mathbf{a}_i = -\frac{G^2}{c^3}\sum_{j\neq i}\frac{m_j}{r_{ij}^3}\{&\mathbf{v}_{ij}\times(3m_i\vec{\alpha}_i + 4m_j\vec{\alpha}_j)\\ &- 3(\mathbf{n}_{ij}\cdot\mathbf{v}_{ij})\mathbf{n}_{ij}\times(m_i\vec{\alpha}_i + 2m_j\vec{\alpha}_j)\\ &- 6\mathbf{n}_{ij}(\mathbf{n}_{ij}\times\mathbf{v}_{ij})\cdot(m_i\vec{\alpha}_i + m_j\vec{\alpha}_j)\}\end{aligned} \tag{9.36}$$

and the rate of change of $\vec{\alpha}$ is

$$\dot{\vec{\alpha}}_i = \frac{G}{2c^2}\sum_{j\neq i}\frac{m_j}{r_{ij}^3}\left(\mathbf{r}_{ij}\times(3\mathbf{v}_i - 4\mathbf{v}_j)\right)\times\vec{\alpha}_i. \tag{9.37}$$

One may consider a system with only two bodies and evaluate the relative acceleration. Ignoring the indices, writing $\mathbf{x}$ for $\mathbf{x}_{12}$, etc. and $M = m_1$, $m = m_2$ and assuming that only one of the bodies ($m_1 = M$) is a big BH, one may neglect the spin of the smaller body $m_2 = m$, i.e. $\vec{\alpha}_2 = \mathbf{0}$. For simplicity one may use the notation

$$\mathbf{g}_a = \mathbf{v}\times\vec{\alpha}, \quad \mathbf{g}_b = (\mathbf{n}\cdot\mathbf{v})\mathbf{n}\times\vec{\alpha}, \; \mathbf{g}_c = (\mathbf{n}\cdot\mathbf{g}_a)\mathbf{n}, \tag{9.38}$$

in terms of which one can write

$$\mathbf{a} = \mathbf{a}_2 - \mathbf{a}_1 = \frac{M}{c^3r^3}\left((4M + 3m)\mathbf{g}_a - (6M + 3m)\mathbf{g}_b - (6M + 6m)\mathbf{g}_c\right) \tag{9.39}$$

and for the rate of change of $\vec{\alpha}$ one has

$$\dot{\vec{\alpha}} = \frac{m(3m + 4M)}{2c^2r^3(M + m)}(\mathbf{v}\times\mathbf{n})\times\vec{\alpha}, \tag{9.40}$$

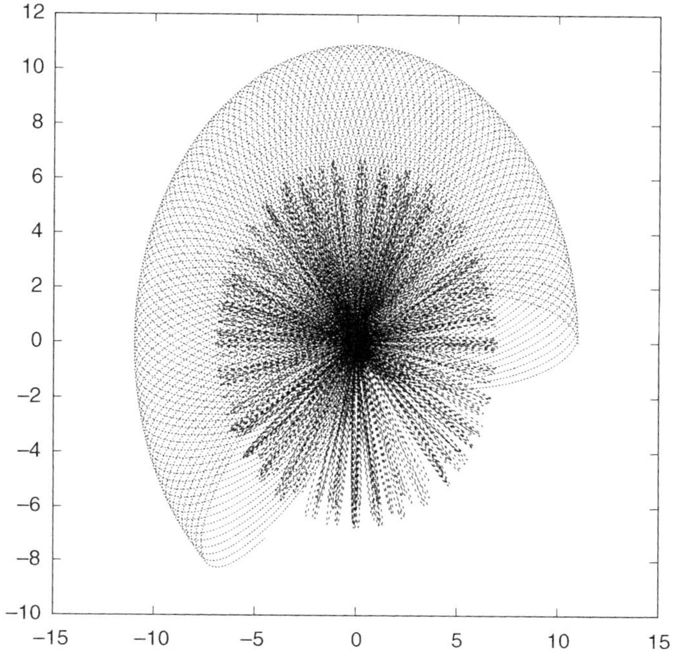

Fig. 9.5. In the field of a rotating black hole, a Kerr hole, the orbits of the stars precess. Both the pericentre's and orbital planes move fast if the orbits are close to the black hole.

in which for velocities in (9.37) the (Newtonian) centre-of-mass velocities have been used.

Figure 9.5 attempts to give an idea of how the Kerr hole field affects the motion of the stars that are gravitationally bound to the hole.

Chapter 10
Artificial Satellite Orbits

This chapter discusses the computation of the orbits of artificial satellites in the gravitational field of the Earth (Mikkola, Palmer and Hashida, 2002). The integration method discussed is the one based on the logarithmic Hamiltonian and the use of the high-order leapfrogs of Yoshida (1990). The methods were discussed earlier in Chapter 8. In Figure 10.1 is an illustration of the strong precessional effects of the Earth's gravitational field, which is not spherically symmetrical, but contains extra terms due to the flattening and bumpy internal structure of the Earth. These effects make the exact computation of artificial satellite motion quite complicated.

10.1 Logarithmic Hamiltonian

Here are used the extended phase space coordinates $(t, \mathbf{r}) = (x_0, x_1, x_2, x_3)$ and corresponding momenta $(p_0, \mathbf{p}) = (p_0, p_1, p_2, p_3)$. Here $x_0 = t$ is the time and $x_k, k = 1, 2, 3$, are the Cartesian coordinates, while p_0 is the momentum of time and $p_k, k = 1, 2, 3$, are the canonically conjugate momenta of the coordinates. The kinetic energy is $T = T(\mathbf{p}) = \mathbf{p}^2/2$ and the force function (negative of potential energy) is $U = U(\mathbf{r}, t)$. The Hamiltonian is $H = T - U$. The homogeneous Hamiltonian

$$H_h = T(\mathbf{p}) + p_0 - U(\mathbf{r}, x_0) \tag{10.1}$$

gives the equations of motion as $\dot{p}_k = -\partial H_h/\partial x_k$, $\dot{x}_k = \partial H_h/\partial p_k$ and $\dot{x}_0 = 1$, $\dot{p}_0 = -\partial H_h/\partial x_0 = -\partial H_h/\partial t$ as expected.

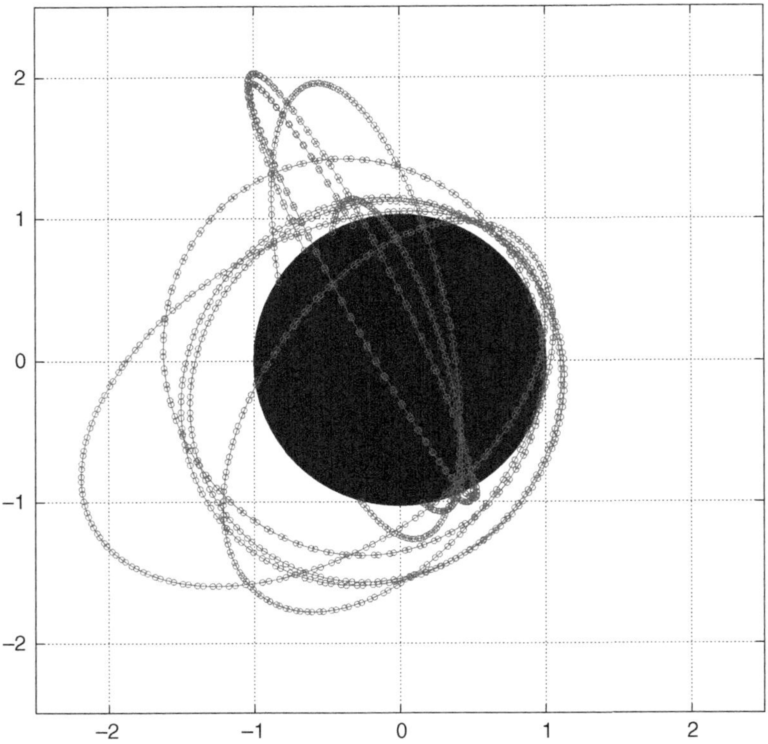

Fig. 10.1. Satellite orbits precess quickly around the rotation axes of the Earth and so does the direction of the pericentre.

Following Mikkola and Tanikawa (1999a) one can take $p_0(0) = -H(0)$ and use the logarithmic Hamiltonian

$$\Lambda = \ln(T(\mathbf{p}) + p_0) - \ln(U(\mathbf{r}, x_0)). \qquad (10.2)$$

This gives the time-transformed equations of motion

$$\mathbf{r}' = \mathbf{p}/T_e, \quad x_0' = 1/T_e, \quad \mathbf{p}' = \frac{\partial U}{\partial \mathbf{r}}/U, \quad p_0' = \frac{\partial U}{\partial x_0}/U, \qquad (10.3)$$

where $T_e = \mathbf{p}^2/2 + p_0$. These are just the normal Hamiltonian equations of motion. The new Hamiltonian Λ does not explicitly depend on the (new) independent variable (call it s). The value of $\Lambda = 0$ along the trajectory and thus similarly $T_e = U$.

Since Λ (10.2) is separable it is possible to apply the leapfrog algorithm

$$\mathbf{L}(h) = \mathbf{X}(h/2)\mathbf{V}(h)\mathbf{X}(h/2), \qquad (10.4)$$

where the operators $\mathbf{X}$ and $\mathbf{V}$ symbolize the x motion

$$\mathbf{X}(s): \quad x_k \leftarrow x_k + s\frac{\partial \Lambda}{\partial p_k} \quad \left(= x_k + \frac{s}{T_e}\frac{\partial T_e}{\partial p_k}\right) \qquad (10.5)$$

and the velocity jump

$$\mathbf{V}(s): \quad p_k \leftarrow p_k - s\frac{\partial \Lambda}{\partial x_k} \quad \left(= p_k + \frac{s}{U}\frac{\partial U}{\partial x_k}\right). \tag{10.6}$$

This algorithm is of second order, but by concatenating leapfrog steps of various lengths (some negative) it is possible to construct higher-order leapfrogs (Yoshida, 1990).

10.2 The Exact Two-Body Leapfrog

When applied to the two-body problem (i.e. $U \propto 1/r$), the leapfrog – and all the higher-order versions of it – give a correct orbital curve but there is a time error (Mikkola and Tanikawa, 1999a; Preto and Tremaine, 1999).

Although this is just another way of writing the two-body solution, it is easily generalized to perturbed motions.

The Keplerian Hamiltonian (in extended phase space) is

$$K = \frac{\mathbf{p}^2}{2} - \frac{M}{r} + p_0, \tag{10.7}$$

and the corresponding logarithmic Hamiltonian reads

$$\Lambda_K = \ln\left(\frac{\mathbf{p}^2}{2} + p_0\right) + \ln(r). \tag{10.8}$$

Here the constant term $(-\ln(M))$ is dropped. This gives the time-transformed equations of motion

$$d\mathbf{r}/ds = \mathbf{p}/T_e, \;\; dt/ds = 1/T_e, \;\; d\mathbf{p}/ds = -\mathbf{r}/r^2, \;\; dp_0/ds = 0, \tag{10.9}$$

where s is the new independent variable. The leapfrog algorithm reads

$$\begin{aligned}
\mathbf{X}(h/2):& \quad \mathbf{r} \leftarrow \mathbf{r} + \frac{h}{2}\mathbf{p}/T_e, \quad t \leftarrow t + \frac{h}{2}\frac{1}{T_e}, \\
\mathbf{V}(h):& \quad \mathbf{p} \leftarrow \mathbf{p} - h\,\mathbf{r}/r^2, \\
\mathbf{X}(h/2):& \quad \mathbf{r} \leftarrow \mathbf{r} + \frac{h}{2}\mathbf{p}/T_e, \quad t \leftarrow t + \frac{h}{2}\frac{1}{T_e}.
\end{aligned} \tag{10.10}$$

By comparison with the exact Kepler motion formulae, Preto and Tremaine (1999) showed that the time error depends only on the stepsize, the mass of the two-body system and the energy, but is independent of eccentricity. In fact one can write the time correction in the form

$$\tau = -\frac{h^3}{4M^2}\mathrm{at}_3\left(\frac{h^2 p_0}{2M^2}\right), \qquad (10.11)$$

where the function $\mathrm{at}_3(\cdot)$ can be defined by $\mathrm{at}_3(x^2) = (x - \mathrm{atan}(x))/x^3$. Adding half of this correction to the increments of time in the above leapfrog formulae gives us an exact algorithm for advancement of the two-body motion.

This is a symplectic algorithm, which one can see after realizing that it is possible to replace the above logarithmic Hamiltonian by

$$\Lambda = \ln(T_e) + \ln(r) + \int_0^{p_0} \frac{\tau}{h}\, dp_0, \qquad (10.12)$$

which gives the same formulae as above, but adds $\tau/2$ to the time increments in (10.10).

Finally it is worth mentioning that the stepsize parameter h is not a step of the independent variable s, but

$$h = \frac{M}{\sqrt{p_0/2}}\tan\left(\frac{h_s}{M}\sqrt{\frac{p_0}{2}}\right) = h_s + \frac{p_0}{6M^2}h_s^3 + O(h_s^5) \qquad (10.13)$$

(where h_s is the s step) and the argument of the tan function is half the increment of the eccentric anomaly, i.e. one may replace the tan above by $\tan(\Delta E/2)$. One notes that, even if h increases without limit, it still gives correct two-body motion, but the maximal advancement of motion is half a period.

10.3 Perturbed Two-Body Motion

10.3.1 Orbit Computation

For perturbed Kepler motion the extended phase space Hamiltonian reads $H = \mathbf{p}^2/2 + p_0 - U(\mathbf{r}, t)$, where

$$U = M/r + R(\mathbf{r}, t). \qquad (10.14)$$

Here R is a small perturbation. The logarithmic Hamiltonian plus time-correction term can be written as

$$\Lambda = \ln(T_e) - \ln(U) + \int_0^{p_0} \frac{\tau}{h}\, dp_0. \qquad (10.15)$$

The leapfrog no longer gives exact solutions; however, one can still add the time-correction term to improve precision provided the perturbation is small.

There are now at least two different ways to proceed further: one can adopt the Wisdom–Holman algorithm, i.e. one divides the Hamiltonian Λ into two integrable parts and uses the composite method (generalized leapfrog); or one may use any of the higher-order leapfrog algorithms of Yoshida (1990). Let us consider first the WH alternative.

WH Method

One can write

$$\Lambda = \ln\left(\frac{\mathbf{p}^2}{2} + p_0\right) + \ln(r) + \int_0^{p_0} \frac{\tau}{h}\, dp_0 - \ln(W), \tag{10.16}$$

where $W = rU$. This formulation shows explicitly the division of Λ into the purely Keplerian part

$$\Lambda_K = \ln\left(\frac{\mathbf{p}^2}{2} + p_0\right) + \ln(r) + \int_0^{p_0} \frac{\tau}{h}\, dp_0 \tag{10.17}$$

and the perturbation part

$$\Lambda_R = -\ln(W). \tag{10.18}$$

Thus the WH method (Wisdom and Holman, 1991) is at our disposal in a time-transformed form (Mikkola, 1997; Rauch and Holman, 1999). The motion due to the Λ_K part can be evaluated with the exact leapfrog formulae $\mathbf{L}(h)$ (with h calculated using (10.13)), while the momentum jumps due to the perturbation Λ_R employ the s step h_s,

$$\Delta\mathbf{p} = -h_s \frac{\partial \Lambda_R}{\partial \mathbf{r}} = h_s \frac{\partial W}{\partial \mathbf{r}} / W, \tag{10.19}$$

$$\Delta p_0 = -h_s \frac{\partial \Lambda_R}{\partial t} = h_s \frac{\partial W}{\partial t} / W. \tag{10.20}$$

Here Δ denotes the jump to be added to the quantity in question.

The time correction in connection with this method requires a comment. Since the mass M does not appear explicitly (but only implicitly in the value of p_0) in the equations of motion (10.9), it cannot appear in the exact solution either. Thus the mass used in the time correction τ must be computed from $M = r(\mathbf{p}^2/2 + p_0)$. Note that, due to the perturbations, this effective value of mass is not exactly the same as the true mass, not even in the limit of very high accuracy. However, as an approximation, it is possible to use the true mass in the time correction and the method is still symplectic.

Y90 Method(s)

A simpler alternative is to use directly the higher-order leapfrog algorithms of Yoshida (1990). If $\mathbf{L}(h)$ means a leapfrog step of length h, then the higher-order leapfrogs may be symbolized by

$$\mathbf{L}_Y(h) = \prod_k \mathbf{L}(d_k h) = \prod_k \left[\mathbf{X}(d_k h/2)\mathbf{V}(d_k h)\mathbf{X}(d_k h/2)\right]. \qquad (10.21)$$

The individual leapfrog (sub)steps can be written

$$\begin{aligned}
\mathbf{X}(d_k h/2):& \quad \mathbf{r} \leftarrow \mathbf{r} + \frac{d_k h}{2}\mathbf{p}/T_e, \quad t \leftarrow t + \frac{d_k h}{2}\frac{1}{T_e} + \frac{1}{2}\tau(d_k h, p_0), \\
\mathbf{V}(d_k h):& \quad \mathbf{p} \leftarrow \mathbf{p} + d_k h\,\frac{\partial U}{\partial \mathbf{r}}, \quad p_0 \leftarrow p_0 + d_k h \frac{\partial U}{\partial t}, \\
\mathbf{X}(d_k h/2):& \quad \mathbf{r} \leftarrow \mathbf{r} + \frac{d_k h}{2}\mathbf{p}/T_e, \quad t \leftarrow t + \frac{d_k h}{2}\frac{1}{T_e} + \frac{1}{2}\tau(d_k h, p_0).
\end{aligned} \qquad (10.22)$$

One should note that the last coordinate motion of step k and the first of step $k+1$ can be combined into a single step of length $(d_k + d_{k+1})h/2$. However, this is not possible for the time correction τ due to the non-linear dependence on the substep length. For the numerical values of the coefficients d_k, see Section 10.5 where they are listed for the fourth-order and sixth-order methods.

10.3.2 State Transition Matrix

In orbit determination from observations one needs to know the state transition matrix $\mathbf{S}(t)$ such that the variations $\Delta\mathbf{r}$, $\Delta\mathbf{p}$ at a given moment of time t can be obtained as

$$\begin{pmatrix}\Delta\mathbf{r}\\ \Delta\mathbf{p}\end{pmatrix} = \mathbf{S}(t)\begin{pmatrix}\Delta\mathbf{r}_0\\ \Delta\mathbf{p}_0\end{pmatrix}, \qquad (10.23)$$

in terms of the initial values $\Delta\mathbf{r}_0$, $\Delta\mathbf{p}_0$. Clearly the columns of this 6×6 matrix can be obtained by solving the variational equations of the system with initial values equal to one of the six-dimensional unit vectors with one of the components $= 1$ and others $= 0$. One can find the variations – the tangent map, as it is also known – in the time-transformed system simply by differentiating the algorithm used to evaluate the motion. In this system there are more variables than in the physical system (namely p_0 and also the time t, which is a coordinate and thus has a variation).

However, one really needs this map in the physical system ($\Delta\mathbf{p}$, $\Delta\mathbf{r}$; t) instead of the time-transformed one ($\delta\mathbf{p}$, δp_0, $\delta\mathbf{r}$, δt; s). The transformation from the latter system to the physical one can be done with

$$\Delta f = \delta f - \delta t \dot{f}, \qquad (10.24)$$

where thus Δf is the physical-space variation of the quantity f and δf is the variation in the time-transformed system, and dot denotes the time derivative. This result follows immediately from the fact that the effect of the variation of the time must be subtracted away to get the result in a system where time does not have variation (= the physical system). Following Mikkola and Innanen (1999) one now computes the tangent map simply by differentiating the algorithm, equation by equation, starting with the initial values for the extended phase space variables given by $\delta\mathbf{r}(0) = \Delta\mathbf{r}(0)$, $\delta\mathbf{p}(0) = \Delta\mathbf{p}(0)$, $\delta p_0 = -\mathbf{p}(0)\cdot\delta\mathbf{p}(0) + \partial U/\partial\mathbf{r}\cdot\delta\mathbf{r}(0)$, $\delta t(0) = 0$. Augmented with the differentials the $\mathbf{X}(s)$ routine reads

$\mathbf{X}(s):$

$$T_e \leftarrow \mathbf{p}^2/2 + p_0, \tag{10.25}$$

$$\delta T_e \leftarrow \mathbf{p}\cdot\delta\mathbf{p} + \delta p_0, \tag{10.26}$$

$$x_0 \leftarrow x_0 + \frac{s}{T_e} + \frac{s}{h}\tau(h, p_0), \tag{10.27}$$

$$\delta x_0 \leftarrow \delta x_0 - \frac{s}{T_e^2}\delta T_e + \frac{s}{h}\frac{\partial\tau(h, p_0)}{\partial p_0}\delta p_0, \tag{10.28}$$

$$x_k \leftarrow x_k + \frac{s}{T_e}p_k, \tag{10.29}$$

$$\delta x_k \leftarrow \delta x_k + \frac{s}{T_e}\delta p_k - \frac{s}{T_e^2}p_k\delta T_e, \tag{10.30}$$

$k = 1, 2, 3,$

and the $\mathbf{V}(s)$ routine may be written

$\mathbf{V}(s):$

$$p_k \leftarrow p_k + s\frac{\partial \ln(U)}{\partial x_k}, \tag{10.31}$$

$$\delta p_k \leftarrow \delta p_k + s\sum_{j=0}^{3}\frac{\partial^2\ln(U)}{\partial x_k\partial x_j}\delta x_j, \tag{10.32}$$

$k = 0, 1, 2, 3.$

These equations are rather easy to program; however, an expression for the derivative of τ with respect to p_0 is needed. From the expression (10.11) one gets

$$\frac{\partial\tau}{\partial p_0} = -\frac{h^5}{8M^4}\mathrm{at}_3'\left(\frac{h^2 p_0}{2M^2}\right). \tag{10.33}$$

One defines the functions $\mathrm{at}_n(x)$ by

$$\mathrm{atan}(x) = x\,\mathrm{at}_1(x^2) = x - x^3\,\mathrm{at}_3(x^2) = x - \frac{x^3}{3} + x^5\,\mathrm{at}_5(x^2) = \ldots \tag{10.34}$$

or by the recursion

$$\mathrm{at}_n(z) = \frac{1}{n} - z\,\mathrm{at}_{n+2}(z), \tag{10.35}$$

where z is written for x^2. From the definition (10.34) it is possible to derive

$$\mathrm{at}_3'(z) = \frac{3}{2}\,\mathrm{at}_5(z) - \frac{1}{2(1+z)}. \tag{10.36}$$

Since also

$$\mathrm{at}_3(z) = \tfrac{1}{3} - z\mathrm{at}_5(z), \tag{10.37}$$

one sees that both $\mathrm{at}_3(z)$ and its derivative are available if one can compute $\mathrm{at}_5(z)$. For small argument values the power series $\mathrm{at}_5 = 1/5 - z/7 + z^2/9 - z^3/11 + - \cdots$ is useful, but more accurate results can be obtained using rational approximations (see Section 10.5).

10.4 Application to Artificial Satellites

10.4.1 Earth's Potential

The potential of the Earth, or any planet, can be written

$$U = \sum_{n=0}^{N_x} \sum_{m=0}^{n} \frac{P_n^m(\cos(\theta))}{r^{n+1}} (C_n^m \cos(m\psi) + S_n^m \sin(m\psi)), \tag{10.38}$$

where $\psi = \phi - n_E t - \eta_0$ and n_E is Earth's rotational frequency and η_0 initial phase, while N_x defines the order up to which terms are taken into account. The functions P_n^m are the well-known associated Legendre polynomials.

Our algorithm needs the value of this potential, its gradient and also the second derivatives with respect to all the coordinates $(x_0, x_1, x_2, x_3) = (t, \mathbf{r})$. Since the potential is available in terms of the spherical coordinates r, θ, ψ, it is necessary to compute the function W and its derivatives with respect to these coordinates.

For convenience one may write $\ln(U) = -\ln(r) + \ln(rU)$; then for

$$W = rU$$

one has

$$W = \sum_{n=0}^{N_x} \sum_{m=0}^{n} P_n^m(\cos(\theta)) a_n^m(\psi) \left(\frac{1}{r}\right)^n, \tag{10.39}$$

where thus

$$a_n^m(\psi) = (C_n^m \cos(m\psi) + S_n^m \sin(m\psi)). \tag{10.40}$$

Using the starting values

$$P_m^m(c) = (2m-1)!!\, s^m, \quad P_{m+1}^m = (2m+1)\, c\, P_m^m(c), \tag{10.41}$$

where $c = \cos(\theta)$, $s = \sin(\theta)$ for short, the associated Legendre polynomials for $n = m+2, \ldots, N_x$ can be obtained from the relation

$$P_n^m = c\alpha_n^m P_{n-1}^m + \beta_n^m P_{n-2}^m, \tag{10.42}$$

where

$$\alpha_n^m = \frac{2n+1}{n-m}, \quad \beta_n^m = -\frac{n-1+m}{n-m}, \tag{10.43}$$

while the first and second derivatives of P_n^m (which are needed in the computation of the forces and the state transition matrix) can be calculated by

$$\frac{dP_n^m}{d\theta} = m\frac{c}{s}P_n^m - P_n^{m+1}, \tag{10.44}$$

$$\frac{d^2P_n^m}{d\theta^2} = m\left(-\frac{1}{s^2}P_n^m + \frac{c}{s}\frac{dP_n^m}{d\theta}\right) - \frac{dP_n^{m+1}}{d\theta}. \tag{10.45}$$

In these formulae one may use a definition of P_n^m which excludes a factor of $(-1)^m$.

The values of $c_m = \cos(m\psi)$ and $s_m = \sin(m\psi)$, in addition to the readily available (see below) c_1, s_1 and $c_0 = 1$, $s_0 = 0$, are obtained by the well-known relations for $m = 2, \ldots, N_x$:

$$c_m = 2c_1c_{m-1} - c_{m-2}, \tag{10.46}$$

$$s_m = 2c_1s_{m-1} - s_{m-2}. \tag{10.47}$$

The input variables needed are the time and coordinates $(t, \mathbf{r}) = (x_0, x_1, x_2, x_3)$ and the initial phase of the Earth's rotation angle η_0 at $t = 0$. Knowing these it is possible to calculate:

$$\rho = \sqrt{x_1^2 + x_2^2}, \tag{10.48}$$

$$r = \sqrt{x_1^2 + x_2^2 + x_3^2}, \tag{10.49}$$

$$c = z/r, \tag{10.50}$$

$$s = \rho/r, \tag{10.51}$$

$$\eta = n_E x_0 + \eta_0, \quad (\psi = \phi - \eta), \tag{10.52}$$

$$c_1 = \cos(\psi) = (x_1\cos(\eta) + x_2\sin(\eta))/\rho, \tag{10.53}$$

$$s_1 = \sin(\psi) = (x_2\cos(\eta) - x_1\sin(\eta))/\rho. \tag{10.54}$$

One should note that, although there are numerous trigonometric functions in the expressions above, only two need to be evaluated, namely $\cos(\eta)$ and $\sin(\eta)$, which define the phase of the Earth. All the rest can be expressed in terms of the coordinates.

The function W can be seen as a polynomial in $1/r$:

$$W = \sum_{n=0}^{N_x} \sigma_n(\theta, \psi) \left(\frac{1}{r}\right)^n, \quad \sigma_n(\theta, \psi) = \sum_{m=0}^{n} P_n^m(c) a_n^m(\psi). \tag{10.55}$$

Thus the standard algorithm to evaluate polynomials,

$$W \leftarrow 0 (\text{initialization}), \tag{10.56}$$

$$W \leftarrow \sigma_k(\theta, \psi) + \frac{1}{r} W, \quad k = n, n-1, \ldots, 0, \tag{10.57}$$

can be used to obtain the value of W. Expressions for derivative evaluations can be obtained simply by differentiating the algorithm(s) needed for evaluation of W. Using bracketed superscripts for derivatives the entire computation can be written as follows:

Initialize W and all its partial derivatives to zero;
then for $n = N_x, N_x - 1, \ldots, 0$ do

$$\sigma_n \leftarrow 0, \quad \sigma_n^{(\psi)} \leftarrow 0, \sigma_n^{(\psi\psi)} \leftarrow 0, \sigma_n^{(\theta)} \leftarrow 0, \sigma_n^{(\theta\theta)} \leftarrow 0, \sigma_n^{(\theta\psi)} \leftarrow 0$$

for $m = 0, \ldots, n$ do

$$a_n^m \leftarrow C_n^m c_m + S_n^m s_m, \tag{10.58}$$

$$a_n^{m(\psi)} \leftarrow m(-C_n^m s_m + S_n^m c_m), \tag{10.59}$$

$$a_n^{m(\psi\psi)} \leftarrow -m^2 a_n^m, \tag{10.60}$$

$$\sigma_n \leftarrow \sigma_n + P_n^m a_n^m, \tag{10.61}$$

$$\sigma_n^{(\psi)} \leftarrow \sigma_n^{(\psi)} + P_n^m a_n^{m(\psi)}, \tag{10.62}$$

$$\sigma_n^{(\psi\psi)} \leftarrow \sigma_n^{(\psi\psi)} + P_n^m a_n^{m(\psi\psi)}, \tag{10.63}$$

$$\sigma_n^{(\theta)} \leftarrow \sigma_n^{(\theta)} + P_n^{m(\theta)} a_n^m, \tag{10.64}$$

$$\sigma_n^{(\theta\theta)} \leftarrow \sigma_n^{(\theta\theta)} + P_n^{m(\theta\theta)} a_n^m, \tag{10.65}$$

$$\sigma_n^{(\theta\psi)} \leftarrow \sigma_n^{(\theta\psi)} + P_n^{m(\theta)} a_n^{m(\psi)}. \tag{10.66}$$

Now σ_n and the derivatives are available for

$$W^{(rr)} \leftarrow \frac{1}{r}\left(W^{(rr)} + 2\frac{1}{r}\left(-W^{(r)} + \frac{1}{r} W\right)\right), \tag{10.67}$$

$$W^{(r\theta)} \leftarrow \frac{1}{r}\left(W^{(r\theta)} - \frac{1}{r} W^{(\theta)}\right), \tag{10.68}$$

$$W^{(r\psi)} \leftarrow \frac{1}{r}\left(W^{(r\psi)} - \frac{1}{r}\,W^{(\psi)}\right), \tag{10.69}$$

$$W^{(\psi\psi)} \leftarrow \sigma_n^{(\psi\psi)} + \frac{1}{r}\,W^{(\psi\psi)}, \tag{10.70}$$

$$W^{(\theta\theta)} \leftarrow \sigma_n^{(\theta\theta)} + \frac{1}{r}\,W^{(\theta\theta)}, \tag{10.71}$$

$$W^{(\theta\psi)} \leftarrow \sigma_n^{(\theta\psi)} + \frac{1}{r}\,W^{(\theta\psi)}, \tag{10.72}$$

$$W^{(\psi)} \leftarrow \sigma_n^{(\psi)} + \frac{1}{r}\,W^{(\psi)}, \tag{10.73}$$

$$W^{(\theta)} \leftarrow \sigma_n^{(\theta)} + \frac{1}{r}\,W^{(\theta)}, \tag{10.74}$$

$$W^{(r)} \leftarrow \frac{1}{r}\left(W^{(r)} - \frac{1}{r}\,W\right), \tag{10.75}$$

$$W \leftarrow \sigma_n + \frac{1}{r}\,W, \tag{10.76}$$

which gives the values of the quantities when n reaches 0. Note that on the right-hand side of the above expressions one must use the 'old' values of the quantities. Therefore the order of evaluation of the updates must be: first the second derivatives then the first derivatives and finally W.

To get the derivatives of the logarithmic Hamiltonian with respect to the coordinates (x_0, x_1, x_2, x_3), further transformations are required. For the first partial derivatives

$$f_k = -\Lambda^{(x_k)} = \partial \ln(U)/\partial x_k = -x_k/r^2 + W^{(x_k)}/W \tag{10.77}$$

one needs the derivatives $W^{(x_k)}$. Remembering that $\psi = \phi - n_E x_0 + \eta_0$ one obtains

$$W^{(x_0)} = -n_E\,W^{(\psi)}, \tag{10.78}$$

$$W^{(x_1)} = W^{(r)}x_1/r + W^{(\theta)}\theta^{(x_1)} + W^{(\psi)}\phi^{(x_1)}, \tag{10.79}$$

$$W^{(x_2)} = W^{(r)}x_2/r + W^{(\theta)}\theta^{(x_2)} + W^{(\psi)}\phi^{(x_2)}, \tag{10.80}$$

$$W^{(x_3)} = W^{(r)}x_3/r + W^{(\theta)}\theta^{(x_3)} + W^{(\psi)}\phi^{(x_3)}, \tag{10.81}$$

and the derivatives of the angles may be evaluated as

$$\theta^{(x_1)} = x_1x_3/(\rho r^2),\;\; \theta^{(x_2)} = x_2x_3/(\rho r^2),\;\; \theta^{(x_3)} = -\rho/r^2, \tag{10.82}$$

$$\psi^{(x_1)} = -x_2/\rho^2, \psi^{(x_2)} = x_1/\rho^2. \psi^{(x_3)} = 0. \tag{10.83}$$

With the above computation of the derivatives $-\Lambda^{(x_i x_j)} = \partial f_i / \partial x_j$ is straightforward. It is $\ln(U) = -\ln(r) + \ln(W)$ and thus the contributions from the simple first part are

$$(-\ln(r))^{(x_i x_j)} = -\frac{\delta_{ij}}{r^2} + 2\frac{x_i x_j}{r^4}. \tag{10.84}$$

Writing $\mathbf{y} = (y_1, y_2, y_3) = (r, \theta, \psi)$ one gets for the second part

$$W^{(x_i x_j)} = \sum_{\alpha\beta} W^{(y_\alpha y_\beta)} y_\alpha^{(x_i)} y_\beta^{(x_j)} + \sum_k W^{(y_k)} y_k^{(x_i x_j)}, \tag{10.85}$$

and the final evaluation of the derivatives of $\ln(W)$ is given by

$$(\ln(W))^{(x_i x_j)} = \frac{W^{(x_i x_j)}}{W} - \frac{W^{(x_i)} W^{(x_j)}}{W^2}. \tag{10.86}$$

10.4.2 Air Drag and Other Effects

The inclusion of the drag effect into a symplectic algorithm is not strictly speaking possible because the system is no longer Hamiltonian. However, it is possible to use the composite method approach (Mikkola, 1998).

It is possible to include the effect in between the symplectic integration steps – which only use the gravitational force – as follows. The equations of motion with drag for $\mathbf{p}$ are

$$\mathbf{p}' = \frac{\partial U}{\partial \mathbf{r}} / U - Cp\mathbf{p}/U, \tag{10.87}$$

where $p = |\mathbf{p}|$ and C is the drag coefficient proportional to the air density and satellite's cross-sectional area and inversely proportional to satellite's mass. The first part of this is taken into account in the symplectic integration step with the Y90 method. The drag part is taken separately into account by solving

$$\mathbf{p}' = -\epsilon p\, \mathbf{p}, \tag{10.88}$$

where $\epsilon = C/U$ has been introduced for short. When solved over a step of length h, keeping the position constant, one gets the approximation

$$\Delta\mathbf{p} = -\epsilon h p_i \mathbf{p}_i / (1 + \epsilon h p_i), \tag{10.89}$$

$$\mathbf{p} \leftarrow \mathbf{p} + \Delta\mathbf{p}, \tag{10.90}$$

where $\mathbf{p}_i$ is the initial value of $\mathbf{p}$. Since the air drag is usually known only approximately, it is not necessary to evaluate the entire function U but it is sufficient to replace it by the leading term $U \approx M/r$, which simplifies the computation.

When the momenta are changed like here, one must take into account that p_0 also changes. In fact it is necessary to maintain the relation $\mathbf{p}^2/2+p_0 = U$. In this operation U remains strictly constant and thus one has

$$p_0 \leftarrow p_0 - \Delta\mathbf{p} \cdot (\mathbf{p}_i + \Delta\mathbf{p}/2), \tag{10.91}$$

which takes into account the second-order effect also. The contributions to the variations can be included by the formulae

$$\delta\mathbf{p} \leftarrow \delta\mathbf{p} + \frac{\partial \Delta\mathbf{p}}{\partial \mathbf{p}_i} \cdot \delta\mathbf{p} + \frac{\partial \Delta\mathbf{p}}{\partial \mathbf{r}} \cdot \delta\mathbf{r}, \tag{10.92}$$

$$\delta p_0 \leftarrow \delta p_0 + \delta\mathbf{p}_i \cdot \mathbf{p}_i - \delta\mathbf{p} \cdot \mathbf{p}. \tag{10.93}$$

The expression for δp_0 follows from the fact that it is possible to write $\Delta p_0 = \frac{1}{2}(\mathbf{p}_i^2 - \mathbf{p}^2)$. When one denotes the operations from (10.89) to (10.93) by the symbol $\mathbf{D}(h)$, one can write the algorithm with drag in the form

$$\begin{aligned} &\prod [\mathbf{D}(h/2)\mathbf{L}_Y(h)\mathbf{D}(h/2)] \\ &\quad = \mathbf{D}(h/2)\mathbf{L}_Y(h)\mathbf{D}(h)\mathbf{L}_Y(h)\mathbf{D}(h) \cdots \mathbf{L}_Y(h)\mathbf{D}(h/2). \end{aligned} \tag{10.94}$$

Thus it is possible to include the drag effect by using a trapezoidal-like rule. Normally this effect is very small and therefore this approximation is sufficiently accurate.

Other effects, such as lunisolar terms and radiation pressure, may be included in the same way as the air drag: one simply adds the term into equation (10.88). Assuming a term $\Delta\mathbf{f}(\mathbf{r}, t)$ has been added to the acceleration, instead of (10.88) one has

$$\mathbf{p}' = -\epsilon p \mathbf{p} + \Delta\mathbf{f}/U, \tag{10.95}$$

which is difficult to solve, but may be approximated by the simple algorithm

$$\mathbf{p} \leftarrow \mathbf{p} + \frac{h}{2}\Delta\mathbf{f}/U, \tag{10.96}$$

$$\mathbf{p} \leftarrow \mathbf{p} - \epsilon h p \mathbf{p}/(1 + \epsilon h p), \tag{10.97}$$

$$\mathbf{p} \leftarrow \mathbf{p} + \frac{h}{2}\Delta\mathbf{f}/U, \tag{10.98}$$

instead of (10.89). If the additional terms are small – as they usually are – this approximation is sufficient for most purposes. The jump of the momentum of time p_0 must be made in the same way as above in (10.91) using the total change of the momentum $\mathbf{p}$.

The modifications for the advancement of the differentials δp_0 and $\delta\mathbf{p}$ can be readily obtained from the above equations.

To obtain the air drag coefficient, so far a simple average table of air density as a function of altitude has been used. One may interpolate linearly the logarithm of the density, which implies the assumption that the density behaves exponentially in between the tabular points.

10.4.3 Checking the Computation

In addition to the conservation of the Hamiltonian (or the logarithmic version of it, or the time-corrected one (10.12)) the differential of the Hamiltonian

$$\delta\Lambda = \frac{\partial\Lambda}{\partial\mathbf{p}}\cdot\delta\mathbf{p} + \frac{\partial\Lambda}{\partial p_0}\delta p_0 + \frac{\partial\Lambda}{\partial\mathbf{r}}\cdot\delta\mathbf{r} + \frac{\partial\Lambda}{\partial t}\delta t, \tag{10.99}$$

where the δ are the variations, should remain constant to high accuracy. If the external perturbations, air drag and other possible forces, are taken into account as described above, this test still applies due to the way the change of p_0 is calculated in (10.91). Unfortunately this also means that we do not have any simple way to check for the correctness of the addition of the extra terms.

10.5 Methods and Functions

10.5.1 Fourth- and Sixth-Order Leapfrogs

The higher-order leapfrog methods (Yoshida, 1990) may be written

$$\begin{aligned}\mathbf{L}_Y(h) &= \prod_{k=1}^{N}\mathbf{L}(d_k h) = \prod_{k=1}^{N}[\mathbf{X}(d_k h/2)\mathbf{V}(d_k h)\mathbf{X}(d_k h/2)]\\ &= \mathbf{X}(d_1 h/2)\left[\prod_{k=1}^{N-1}\mathbf{V}(d_k h)\mathbf{X}((d_k + d_{k+1})h/2)\right]\\ &\quad \mathbf{V}(d_N h)\mathbf{X}(d_N h/2).\end{aligned}$$

The number of function evaluations N, the one-step error (order) err_1 and the substep lengths d_k are as follows.

1. In the fourth-order method

$$N = 3, \quad err_1 = O(h^5)$$

and the values of the parameters d_k are

$$\{d_1, d_2, d_3\} = \left\{\frac{1}{2-2^{1/3}}, \frac{-2^{1/3}}{2-2^{1/3}}, \frac{1}{2-2^{1/3}}\right\}.$$

2. In the sixth-order method

$$N = 7, \quad err_1 = O(h^7)$$

and the d_k as given in Yoshida (1990) read

$$\{d_1, d_2, d_3, d_4, d_5, d_6, d_7\} = \{0.7845136104775599, 0.2355732133593569, \\ -1.17767998417887, 1.315186320683906, \\ -1.17767998417887, 0.2355732133593569, \\ 0.7845136104775599\}.$$

10.5.2 Auxiliary Functions

In the computation of the time correction the special functions

$$\mathrm{at}_n(z) = \frac{1}{n} - \frac{z}{n+2} + \frac{z^2}{n+4} - \frac{z^3}{n+6} + - \cdots \tag{10.100}$$

are useful. These are related to the arctangent function by

$$\mathrm{atan}(x) = x\,\mathrm{at}_1(x^2) = x - x^3\,\mathrm{at}_3(x^2) = x - \frac{x^3}{3} + x^5\,\mathrm{at}_5(x^2) = \cdots. \tag{10.101}$$

These functions (or rather at_3 and/or at_5) are usually needed at small argument value (in practical use it often is $z < 10^{-2}$) and therefore the expressions in terms of the arctangent are not very useful. With an argument value as small as in our experiments, power series may be satisfactory. One may also use a Padé approximant for this function.

If it is desirable to have accurate results for all argument values, which may be useful for testing one's code or for hyperbolic orbits, the following can be used. For $\mathrm{at}_5(x)$ the following Padé approximant can be derived:

$$P(z) = 1 + \frac{569z}{203} + \frac{5350z^2}{1827} + \frac{28078z^3}{20097} + \frac{5325469z^4}{18027009} \\ + \frac{5708201z^5}{270405135} + \frac{1048576z^6}{203795336745}, \tag{10.102}$$

$$Q(z) = 5 + \frac{510z}{29} + \frac{2125z^2}{87} + \frac{4420z^3}{261} + \frac{12155z^4}{2001} \\ + \frac{4862z^5}{4669} + \frac{2431z^6}{38019}, \tag{10.103}$$

$$\mathrm{at}_5(z) \approx P(z)/Q(z) \quad (-2.5 \times 10^{-9} z^{13} + \cdots) \\ \text{(valid for } |z| < 0.25). \tag{10.104}$$

Using the formula for the tangent of a double angle it is possible to deduce formulae for evaluating the at_5 function at any value of the argument. The following can be used to reduce the argument size in computation of $\mathrm{at}_5(z)$:

$$K = 1 + \sqrt{1+z}, \tag{10.105}$$

$$X = z/K^2, \tag{10.106}$$

$$\mathrm{at}_5(z) = \frac{1}{3K^5}(K(1+K)^2 + 6\,\mathrm{at}_5(X)). \tag{10.107}$$

With the above expression for $\mathrm{at}_5(z)$ one needs to use this reduction only once (i.e. one substitutes the Padé approximant value for at_5 in the right-hand side of the above formula) for elliptical orbits (or more generally for $z > -0.7$) to get a near-double-precision result.

For the computation of the derivative of the at_n functions one easily obtains (by induction) the following:

$$\mathrm{at}'_n(z) = \frac{n}{2}\,\mathrm{at}_{n+2}(z) - \frac{1}{2(1+z)}.$$

The functions at_3 and at'_3 can then be obtained by

$$\mathrm{at}_3(z) = \frac{1}{3} - z\,\mathrm{at}_5(z), \tag{10.108}$$

$$\mathrm{at}'_3(z) = \frac{3}{2}\,\mathrm{at}_5(z) - \frac{1}{2(1+z)}. \tag{10.109}$$

References

Aarseth, S. J. (2010). *Gravitational N-Body Simulations*. Cambridge, UK: Cambridge University Press.

Aarseth, S. J. and Zare, K. (1974). A regularization of the three-body problem. *Celestial Mechanics*, **10**, 185–205.

Blanchet, L. and Iyer, B. R. (2003). Third post-Newtonian dynamics of compact binaries: equations of motion in the centre-of-mass frame. *Classical and Quantum Gravity*, **20**, 755–776.

Bulirsch, R. and Stoer, J. (1966). Numerical treatment of ordinary differential equations by extrapolation methods. *Numerische Mathematik*, **8**, 1–13.

Burkardt, T. M. and Danby, J. M. A. (1983). The solution of Kepler's equation, II. *Celestial Mechanics*, **31**, 317–328.

Burrau, C. (1913). Numerische Berechnung eines Spezialfalles des Dreikörperproblems. *Astronomische Nachrichten*, **195**, 113–116.

Chambers, J. (1999). A hybrid symplectic integrator that permits close encounters between massive bodies. *Monthly Notices of the Royal Astronomical Society*, **304**, 793–799.

Chambers, J. E. and Murison, M. A. (2000). Pseudo-high-order symplectic integrators. *The Astronomical Journal*, **119**, 425–433.

Choi, J. S. and Tapley, B. D. (1973). An extended canonical perturbation method. *Celestial Mechanics*, **7**, 77–90.

Conway, B. A. (1986). An improved algorithm due to Laguerre for the solution of Kepler's equation. *Celestial Mechanics*, **39**, 199–211.

Cordeiro, R. R., Gomes, R. S. and Martins, R. V. (1997). A mapping for nonconservative systems. *Celestial Mechanics and Dynamical Astronomy*, **65**, 407–419.

Danby, J. M. A. (1992). *Fundamentals of Celestial Mechanics*, 2nd ed. Richmond, VA: Willmann-Bell.

Danby, J. M. A. and Burkardt, T. M. (1983). The solution of Kepler's equation, I. *Celestial Mechanics*, **31**, 95–107.

Dey, L., Valtonen, M. J., Gopakumar, A., et al. (2018). Authenticating the presence of a relativistic massive black hole binary in OJ 287 using its general relativity centenary flare: improved orbital parameters. *The Astrophysical Journal*, **866**, 11.

Duncan, M. J., Levison, H. F. and Lee, M. H. (1998). A multiple timestep symplectic algorithm for integrating close encounters. *The Astronomical Journal*, **116**, 2067–2077.

Fukushima, T. (1996). A fast procedure solving Kepler's equation for elliptic case. *The Astronomical Journal*, **112**, 2858.

Fukushima, T. (1997a). A method solving Kepler's equation without transcendental function evaluations. *Celestial Mechanics and Dynamical Astronomy*, **66**, 309–319.

Fukushima, T. (1997b). A procedure solving the extended Kepler's equation for the hyperbolic case. *The Astronomical Journal*, **113**, 1920.

Fukushima, T. (1998). A fast procedure solving Gauss' form of Kepler's equation. *Proc. 30th Symp. on Celestial Mechanics*, p. 217.

Fukushima, T. (1999). Fast procedure solving universal Kepler's equation. *Celestial Mechanics and Dynamical Astronomy*, **75**, 201–226.

Fukushima, T. (2008). An orbital element formulation without solving Kepler's equation. *The Astronomical Journal*, **135**, 72–82.

Funato, Y., Hut, P., McMillan, S. and Makino, J. (1996). Time-symmetrized Kustaanheimo–Stiefel regularization. *The Astronomical Journal*, **112**, 1697.

Goldstein, H., Poole, C. P., Jr and Safko, J. L. (2002). *Classical Mechanics*, 3rd ed. San Francisco, CA: Addison-Wesley.

Gragg, W. B. (1964). Repeated extrapolation to the limit in the numerical solution of ordinary differential equations. PhD thesis, University of California, Los Angeles.

Gragg, W. B. (1965). On extrapolation algorithms for ordinary initial value problems. *SIAM Journal on Numerical Analysis*, **2**, 384–403.

Hagihara, Y. (1970). *Dynamical Principles and Transformation Theory*. Cambridge, MA: MIT Press.

Hand, L. N. and Finch, J. D. (2008). *Analytical Mechanics*. Cambridge, UK: Cambridge University Press.

Heggie, D. C. (1974). A global regularisation of the gravitational N-body problem. *Celestial Mechanics*, **10**, 217–241.

Hellström, C. and Mikkola, S. (2009). Universal formulation of quasi-Keplerian motion, and its applications. *New Astronomy*, **14**, 607–614.

Hellström, C. and Mikkola, S. (2010). Explicit algorithmic regularization in the few-body problem for velocity-dependent perturbations. *Celestial Mechanics and Dynamical Astronomy*, **106**, 143–156.

Herrick, S. (1972). *Astrodynamics*, Vol. II. London, UK: Van Nostrand Reinhold.

von Hoerner, S. (1960). Die numerische Integration des n-Körper-Problems für Sternhaufen. I. *Zeitschrift für Astrophysik*, **50**, 184.

von Hoerner, S. (1963). Die numerische Integration des n-Körper-Problems für Sternhaufen. II. *Zeitschrift für Astrophysik*, **57**, 47.

Hopf, H. (1931). Über die Abbildungen der dreidimensionalen Sphäre auf die Kugelfläche. *Mathematische Annalen*, **104**, 637–665. Reprinted: (1964). *Selecta Heinz Hopf*. Berlin, Germany: Springer, pp. 38–63.

Huang, W. and Leimkuhler, B. (1997). The adaptive Verlet method. *SIAM Journal on Scientific Computing*, **18**, 239–256.

Hut, P., Makino, J. and McMillan, S. (1995). Building a better leapfrog. *The Astrophysical Journal*, **443**, L93–L96.

Kamel, A. A. (1971). Lie transformations and the Hamiltonization of non-Hamiltonian systems. *Celestial Mechanics*, **4**, 397–405.

Kinoshita, H., Yoshida, H. and Nakai. H. (1991). Symplectic integrators and their application in dynamical astronomy. *Celestial Mechanics and Dynamical Astronomy*, **50**, 59–71.

Kiseleva, L. G., Eggleton, P. P. and Mikkola, S. (1998). Tidal friction in triple stars. *Monthly Notices of the Royal Astronomical Society*, **300**, 292–302.

Koseleff, P.-V. (1993). Relations among Lie formal series and construction of symplectic integrators. In G. Cohen, T. Mora and O. Moreno, eds, *Applied Algebra, Algebraic Algorithms and Error Correcting Codes, AAECC-10*. New York, NY: Springer, pp. 213–230.

Kustaanheimo, P. (1964). Die Spinordarstellung der energetischen Identitäten der Keplerbewegung. In E. Stiefel, ed., *Mathematische Methoden der Himmelsmechanik und Astronautik*, Mathematisches Forschungsinstitut Oberwolfach, Berichte 1. Mannheim, Germany: Bibliographisches Institut Mannheim, pp. 333–340.

Kustaanheimo, P. and Stiefel, E. (1965). Perturbation theory of Kepler motion based on spinor regularization. *Journal für die Reine und Angewandte Mathematik*, **218**, 204–219.

Laskar, J. and Robutel, P. (2001), High order symplectic integrators for perturbed Hamiltonian systems. *Celestial Mechanics and Dynamical Astronomy*, **80**, 39–62.

Levi-Civita, T. (1920). Sur la régularisation du problème des trois corps. *Acta Mathematica*, **42**, 99–144.

Makino, J. and Aarseth, S. J. (1992). On a Hermite integrator with Ahmad–Cohen scheme for gravitational many-body problems. *Publications of the Astronomical Society of Japan*, **44**, 141–151.

McLachlan, R. I. (1995). Composition methods in the presence of small parameters. *BIT Numerical Mathematics*, **35**(2), 258–268.

Mikkola, S. (1983). Encounters of binaries – I. Equal energies. *Monthly Notices of the Royal Astronomical Society*, **203**, 1107–1121.

Mikkola, S. (1984a). Encounters of binaries. II – Unequal energies. *Monthly Notices of the Royal Astronomical Society*, **207**, 115–126.

Mikkola, S. (1984b). Encounters of binaries. III – Fly-bys. *Monthly Notices of the Royal Astronomical Society*, **208**, 75–82.

Mikkola, S. (1985). A practical and regular formulation of the N-body equations. *Monthly Notices of the Royal Astronomical Society*, **215**, 171–177.

Mikkola, S. (1987). A cubic approximation for Kepler's equation. *Celestial Mechanics*, **40**, 329–334.

Mikkola, S. (1997). Practical symplectic methods with time transformation for the few-body problem. *Celestial Mechanics and Dynamical Astronomy*, **67**, 145–165.

Mikkola, S. (1998). Non-canonical perturbations in symplectic integration. *Celestial Mechanics and Dynamical Astronomy*, **68**, 249–255.

Mikkola, S. and Aarseth, S. J. (1990). A chain regularization method for the few-body problem. *Celestial Mechanics*, **47**, 375–390.

Mikkola, S. and Aarseth, S. J. (1993). An implementation of N-body chain regularization. *Celestial Mechanics and Dynamical Astronomy*, **57**, 439.

Mikkola, S. and Aarseth, S. J. (1996). A slow-down treatment of close binaries. *Celestial Mechanics and Dynamical Astronomy*, **64**, 197–208.

Mikkola, S. and Aarseth, S. J. (1998). An efficient integration method for binaries in N-body simulations. *New Astronomy*, **3**, 309–320.

Mikkola, S. and Aarseth, S. J. (2002). A time-transformed leapfrog scheme. *Celestial Mechanics and Dynamical Astronomy*, **84**, 343–354.

Mikkola, S. and Innanen, K. (1999). Symplectic tangent map for planetary motions. *Celestial Mechanics and Dynamical Astronomy*, **74**, 59–67.

Mikkola, S. and Innanen, K. (2002). Individual accuracy checks for massive bodies and particles in symplectic integration. *The Astronomical Journal*, **124**, 3445–3448.

Mikkola, S. and Merritt, D. (2006). Algorithmic regularization with velocity-dependent forces. *Monthly Notices of the Royal Astronomical Society*, **372**, 219–223.

Mikkola, S. and Merritt, D. (2008). Implementing few-body algorithmic regularization with post-Newtonian terms. *The Astronomical Journal*, **135**, 2398–2405.

Mikkola, S. and Palmer, P. (2001). Simple derivation of symplectic integrators with first order correctors. *Celestial Mechanics and Dynamical Astronomy*, **77**, 305–317.

Mikkola, S. and Tanikawa, K. (1999a). Algorithmic regularization of the few-body problem. *Monthly Notices of the Royal Astronomical Society*, **310**, 745–749.

Mikkola, S. and Tanikawa, K. (1999b). Explicit symplectic algorithms for time-transformed Hamiltonians. *Celestial Mechanics and Dynamical Astronomy*, **74**, 287–295.

Mikkola, S. and Tanikawa, K. (2007). Correlation of macroscopic instability and Lyapunov times in the general three-body problem. *Monthly Notices of the Royal Astronomical Society*, **379**, L21–L24.

Mikkola, S. and Tanikawa, K. (2013a). Implementation of an efficient logarithmic-Hamiltonian three-body code. *New Astronomy*, **20**, 38–41.

Mikkola, S. and Tanikawa, K. (2013b). Regularizing dynamical problems with the symplectic logarithmic Hamiltonian leapfrog. *Monthly Notices of the Royal Astronomical Society*, **430**, 2822–2827.

Mikkola, S. and Wiegert, P. (2002). Regularizing time transformations in symplectic and composite integration. *Celestial Mechanics and Dynamical Astronomy*, **82**, 375–390.

Mikkola, S., Palmer, P. and Hashida, Y. (2002). An implementation of the logarithmic Hamiltonian method for artificial satellite orbit determination. *Celestial Mechanics and Dynamical Astronomy*, **82**, 391–411.

Mikkola, S., Brasser, R., Wiegert, P. and Innanen, K. (2004). Asteroid 2002 VE68, a quasi-satellite of Venus. *Monthly Notices of the Royal Astronomical Society*, **351**, L63–L65.

Mora, T. and Will, C. M. (2004). Post-Newtonian diagnostic of quasiequilibrium binary configurations of compact objects. *Physical Review D*, **69**, 104021.

Poincaré, H. (1895). *Les Méthodes Nouvelles de la Mécanique Céleste*. Paris, France: Gauthier-Villars. Republished in English: (1993). *New Methods in Celestial Mechanics*. New York, NY: AIP Press.

Press, W. H., Flannery, B. P., Teukolsky, S. A. and Vetterling, W. T. (1986). *Numerical Recipes*. Cambridge, UK: Cambridge University Press.

Preto, M. and Tremaine, S. (1999). A class of symplectic integrators with adaptive time step for separable Hamiltonian systems. *The Astronomical Journal*, **118**, 2532–2541.

Rauch, K. P. and Holman, M. (1999). Dynamical chaos in the Wisdom–Holman integrator: origins and solution. *The Astronomical Journal*, **117**, 1087–1102.

Saha, P. and Tremaine, S. D. (1992). Symplectic integrators for Solar System dynamics. *The Astronomical Journal*, **104**, 1633–1640.

Saha, P. and Tremaine, S. D. (1994). Long-term planetary integrations with individual timesteps. *The Astronomical Journal*, **108**, 1962–1969.

Sanz-Serna, J. M. (1992). Symplectic integrators for Hamiltonian problems: an overview. *Acta Numerica*, **1**, 243–286.

Soffel, M. H. (1989). *Relativity in Astrometry, Celestial Mechanics and Geodesy*. Berlin, Germany: Springer, p. 141.

Steffensen, J. F. (1907). Über die Integration des Dreikörperproblems in der Ebene. *Astronomische Nachrichten*, **176**, 221.

Stiefel, E. L. and Scheifele, G. (1971). *Linear and Regular Celestial Mechanics*. Berlin, Germany: Springer.

Stumpff, K. (1959). *Himmelsmechanik*, Band I. Berlin, Germany: VEB Deutscher Verlag der Wissenschaften.

Szebehely, V. and Peters, C. F. (1967). Complete solution of a general problem of three bodies. *The Astronomical Journal*, **72**, 876.

Szebehely, V. and Zare, K. (1975). Time transformations in the extended phase-space. *Celestial Mechanics*, **11**, 469.

Tanikawa, K., Saito, M. M. and Mikkola, S. (2019). A search for triple collision orbits inside the domain of the free-fall three-body problem. *Celestial Mechanics and Dynamical Astronomy*, **131**, 24.

Tremaine, S. (2001). Canonical elements for collision orbits. *Celestial Mechanics and Dynamical Astronomy*, **79**, 231–233.

Valtonen, M. J., Mikkola, S., Merritt, D., et al. (2010). Measuring the spin of the primary black hole in OJ287. *The Astrophysical Journal*, **709**, 725–732.

Waldvogel, J. (1997). Symplectic integrators for Hill's lunar problem. In R. Dvorak and J. Henrard, eds, *The Dynamical Behaviour of Our Planetary System*. Dordrecht, The Netherlands: Kluwer Academic, pp. 291–305.

Waldvogel, J. (2008). Quaternions for regularizing celestial mechanics: the right way. *Celestial Mechanics and Dynamical Astronomy*, **102**, 149–162.

Wisdom, J. and Holman, M. (1991). Symplectic maps for the N-body problem. *The Astronomical Journal*, **102**, 1520–1538.

Wisdom, J., Holman, M. and Touma, J. (1996). Symplectic correctors. In J. E. Marsden, G. W. Patrick and W. F. Shadwick, eds, *Integration Algorithms and Classical Mechanics (Proc. Integration Methods in Classical Mechanics Meeting, Waterloo, ON, 14–18 October 1993)*. Fields Institute Communications, Vol. 10. Providence, RI: American Mathematical Society, p. 217.

Yoshida, H. (1990). Construction of higher order symplectic integrators. *Physics Letters A*, **150**, 262–268.

Zare, K. (1974). A regularization of multiple encounters in gravitational N-body problems. *Celestial Mechanics*, **10**, 207–215.

Index

Accuracy checks in WH methods, 103
Adams, Störmer and Cowell integrators, 58
Air drag, 232
Algorithmic regularization, 177, 189
Alternatives for WH, 119
Angular elements, 21
Artificial satellites, 228
AZ method, 153
AZ method coordinates, 154

Behaviour at event horizon, 215
Binary as a harmonic oscillator, 150
Black hole spins, 217

c-function routine, 33
c-functions, 31, 32, 124
Canonical transformations, 37, 38, 83, 85
Chain, 161
Chain derivatives evaluation, 165
Chain formation, 160
Chain method, 158, 171, 190
Chain updating, 161, 164
Chaos in the Solar System, 102
Comet perturbations, 53
Composite integration, 87
Composite integration methods, 73

Difference methods, linear operators, 55
Different regularized vectors, 147

Earth's atmosphere effect, 232
Earth's potential, 228
Earth's potential gradient, 228
Ellipse with leapfrog, 76
Elliptic two-body motion, 27
Elliptical element perturbations, 51
Energy errors (three-body), 209
Energy errors in leapfrog integration, 76
Equation solution by bracketing, 30
Error Hamiltonian, 88
Error in warm-up, 113
Exact Kepler leapfrog, 132
Extended phase space, 143
External forces, 140, 187, 189, 192
External forces (chain), 162, 166
Extrapolation method, 72

First-order equations by extrapolation, 72
Five-body interaction, 185
Five-body orbit, 6
Fourth-order code, 90
Functional time transformation, 135

Gauss' *f*–*g* functions, 27
Generating function, 38, 85
Geocentric Solar System, 8
Geometry of ellipse, 20
Global regularization, 154
Global regularization algorithm, 156
Global regularization coordinates, 157

Hamiltonian, 36, 39
Hamiltonian equations, 37
Heliocentric coordinate system, 82
Heliocentric coordinates, 8
Herrick's variation, 46
History, 10

Implicit Runge–Kutta–Gaussian integrators, 63
In-spiralling due to GR radiation, 215
Infinitesimal transformations, 83
Initial values as elements, 47
Inner Solar System, 7
Integrals of motion, 5

Jacobian coordinate system, 82
Jacobian coordinates, 9, 95, 99

Kepler's equation, 19, 23, 145
Kepler's equation (*N*-body), 168
Kepler's equation (universal), 28

Kerr hole spins, 217
KS-transformation versus physical, 150
KS-transformation, 124, 144, 146

Lagrange and Poisson equations, 49
Lagrange equations, 35, 43
Lagrangian, 35
Leapfrog, 74, 88, 173
Leapfrog (exact for binary), 223
Leapfrog (generalized), 92, 175
Leapfrog (higher-order), 78, 89, 225
Leapfrog (quasi-Keplerian), 198
Leapfrog (three-body problem), 207
Leapfrog (time-transformed), 185
Leapfrog (two-body), 181
Leapfrog (velocity-dependent force), 194
Leapfrogs (higher-order), 234
Levi-Civita regularization, 145
Lie series, 84
Logarithmic Hamiltonian, 131, 178, 205, 221, 224
Logarithmic leapfrog (two-body), 182
Lyapunov exponents, 94

Modified potential (WH method), 118
Motion and variations, 100

N-body Hamiltonian in KS-formulation, 160
N-body problem, 5

Perturbations of Delaunay variables, 52
Perturbative variation, 46
Picard's iteration, 86
Planetary dynamics, 8
PN-corrected orbital elements, 217
PN-corrected semi-major axis versus stepnumber, 216
PN-corrected semi-major axis versus time, 216
Point transformation, 38
Poisson equations, 45
Polar coordinates, 38
Post-Newtonian terms, 211
Pythagorean three-body orbit, 155

Quasi-Keplerian motion, 199
Quasi-satellite of Venus, 93

Radius of convergence, 69–71
Regions for methods, 12
Relative coordinates (three-body), 206

Satellite dynamics, 9
Satellite orbits, 222
Second-order equations by extrapolation, 73
Sixth-order code, 91
Slow-down of binaries, 170
Solving perturbed harmonic oscillator equations, 151
Spin–orbit terms, 213
Stars in a Kerr hole field, 219
State transition matrix, 226
Steffensen's power series method, 65
Stellar rotation, 204
Stepsize control, 167
Symplectic corrector, 115, 117, 127
Symplectic Euler method, 88
Symplectic integration, 81, 87
Symplectic interpolator, 116
Symplectic tangent map, 96

Taylor series, 17, 56, 65, 84, 152, 176
Taylor series for N-body system, 68
Three-body geometry, 126
Three-body problem, 126, 128, 139, 153, 204
Three-body problem (relative coordinates), 206
Three-body problem with WH, 126
Tidal friction, 203
Time transformation, 17, 41, 125, 128, 131, 135, 143, 167, 178, 193
Time transformation (WH method), 122
Time-transformed Hamiltonian, 122
Time-transformed leapfrog, 185
Transformation to physical variables (KS chain), 163
Transformation to regularized variables (KS chain), 163
Triple collision orbit, 209
Two fixed centres, 137
Two-body (perturbed), 224
Two-body orbit, 5
Two-body power series, 66
Two-body problem, 4, 15
Two-body problem (perturbed), 122, 196
Two-body problem (solution), 16

Units, 3
Universal Kepler's equation, 28, 29
Universal two-body motion, 27

Variation of elements, 50
Variational equations, 77, 97, 99
Variational equations (WH method), 98
Velocity-dependent perturbations, 194

Wisdom–Holman method, 93, 94, 225
Wisdom–Holman with δ-functions, 93

Yoshida's high-order methods, 90
Yoshida's sixth-order method, 90